Introduction to Chemistry for Biology Students

Fifth Edition

George I. Sackheim
Associate Professor Emeritus, Chemistry
University of Illinois at Chicago

formerly
Coordinator of Biological and Physical Sciences
Michael Reese Hospital and Medical Center, Chicago

Science Instructor
St. Francis Hospital School of Nursing
Evanston, Illinois

The Benjamin/Cummings Publishing Company, Inc.

Menlo Park, California • Reading, Massachusetts • New York • Don Mills, Ontario
Wokingham, U.K. • Amsterdam • Bonn • Paris • Milan • Madrid • Sydney
Singapore • Tokyo • Seoul • Taipei • Mexico City • San Juan, Puerto Rico

Associate Editor: *Leslie A. With*
Sponsoring Editor: *Anne Scanlan-Rohrer*
Editorial Assistant: *Sharon Sforza*
Production Editor: *Donna Linden*
Composition and Film Supervisor: *Vivian McDougal*
Design Manager: *Don Kesner*
Permissions Editor: *Marty Granahan*
Manufacturing Supervisor: *Merry Free Osborn*
Marketing Manager: *Nathalie Mainland*
Copy Editor: *Alan Titche*
Proofreader: *Rosemary Sheffield*
Text Designer: *Patrick Devine, Jonathan Peck Typographers*
Illustrators: *Jonathan Peck Typographers and Val Felts*
Compositor: *Jonathan Peck Typographers*
Cover Designer: *Annabelle Ison*
Cover Photograph: *Geoff Tompkinson/Science Photo Library*

Credits: Pages 153 and 202: Figures a and b from W. M. Becker and D. W. Deamer, *The World of the Cell*, 2nd ed. (Menlo Park, CA: Benjamin/Cummings, 1991). © 1991 Benjamin/Cummings Publishing Company; Figures d and e © Irving Geis. Pages 210–213: Appendix figures A, B, C, and D adapted from N. A. Campbell, L. G. Mitchell, and J. B. Reece, *Biology: Concepts and Connections* (Menlo Park, CA: Benjamin/Cummings, 1994). © 1994 Benjamin/Cummings Publishing Company.

Library of Congress Cataloging-in-Publication Data
Sackheim, George I.
Introduction to chemistry for biology students / George I. Sackheim. — 5th ed.
p. cm.
ISBN 0-8053-7706-9
1. Chemistry—Programmed instruction. I. Title.
QD33.S134 1996
540'.7'7—dc20 95-12032

2 3 4 5 6 7 8 9 10—CRW—99 98 97 96

The Benjamin/Cummings Publishing Company, Inc.
2725 Sand Hill Road
Menlo Park, CA 94025

Introduction to Chemistry for Biology Students

Contents

To the Student

Introduction to Chemistry for Biology Students, Fifth Edition, is not an ordinary book. It has been programmed to help you review the basic facts, concepts, and terminology of chemistry that are essential to an understanding of biological phenomena. Today's biology courses place increasing emphasis on the chemical processes that underlie the critical biological functions. This book will help you to understand those processes.

The topics involved are among the most critical and exciting that science will explore in the years ahead. What are the basic chemical processes underlying biological phenomena? What are the essential differences between living and nonliving matter? What are the conditions under which molecules organize into living matter? Can these conditions be duplicated experimentally?

If you have already had a course in chemistry, this programmed book can serve as an effective review of the fundamental concepts. If you have had no previous chemistry, the program will give you the background you need to gain a clear understanding of the biological processes you will be studying.

Do not try to skip around in this book. Start at the beginning and proceed through it at the pace that is most comfortable for you. Under no circumstances should you rush through. Whether it takes you two hours or six hours to complete is of little importance. Your primary objective should be to master the material no matter how long that may take.

The material covered in this book will help you most if you complete it during the first two weeks of your biology course. Having done that, you will be ready to handle the chemical aspects of biology as they come up.

When you have completed the program, you may want to repeat certain material. To simplify this process, use the book's Table of Contents to help you locate specific topics.

If you follow the directions and complete this program, you will learn to:

- recognize elements present in various compounds
- know what is meant by pH and by ionization
- recognize acids, bases, and salts
- discriminate between electrolytes and nonelectrolytes

- understand oxidation and reduction
- know what isotopes are
- recognize various organic functional groups
- differentiate among carbohydrates, fats, and proteins
- understand how enzymes function
- recognize nucleic acids
- understand biologic oxidation, including glycolysis, the Krebs cycle, and the electron transport chain

How to Use This Book

This type of instructional book may be new to you. Its subject matter has been presented as a series of numbered problems. Each builds on information you have learned in the preceding problems. For that reason, it is important not to skip around. The sequence of the problems is important because it is programmed to help you learn more efficiently.

Respond to Every Problem

Some problems present new information; others review material presented earlier. Every problem presents a learning task that requires some response from you.

You may be asked to make any of the following types of responses:

- write an answer in a blank space
- label a diagram
- draw a simple diagram
- select the correct answer from among several alternatives
- write a sentence in answer to a question

When you have written or marked your answer, you will want to find out whether you are correct. Programmed instruction provides you with important feedback by giving you easy access to the answers, which are located in shaded areas on the outer side of each page. *Do not look at the correct answer until after you have marked your own answer.* If you look before answering, you will only impair your own learning process.

Use an Answer Mask

Bound into the back of the book is a sheet of heavy paper that is perforated. Tear off the outer portion of the sheet for use as an answer mask. Here is what to do:

1. As you start working on the first page, a right-hand page, cover the shaded answer column with the answer mask *before* looking at the problems. When you turn the page, shift you answer mask to cover the answer column on the left-hand page before reading the problems on that page. *Be sure you have covered the answers before you read anything.*

2. Each problem number is centered on the page with a shaded rectangle behind it. Read the problem carefully, then record your answer. Make sure you either write each answer or do whatever the directions say. Do not simply think of the answer and then go on. Actually writing or marking your answer reinforces your learning.

3. Now move the answer mask aside to reveal the answer, which you will find aligned with the problem number. (If the main part of the problem runs onto the next page, you will find the answer at the top of that page.)

4. If your answer was correct, move on to the next problem.

5. If your answer was incorrect, reread the problem (if necessary, reread several of the preceding problems) until you understand your error and know why the given answer is correct. Then go on.

When you have completely worked your way through this book, you should have the knowledge of chemistry you need to succeed in your biology courses.

PART I

Inorganic Chemistry

Atomic Structure 1

ELEMENTARY PARTICLES

Atoms are made up of several components. Collectively these components are called the *elementary particles*. We will be discussing the three major elementary particles: *protons, neutrons*, and *electrons*.

Here is a diagram of an atom:

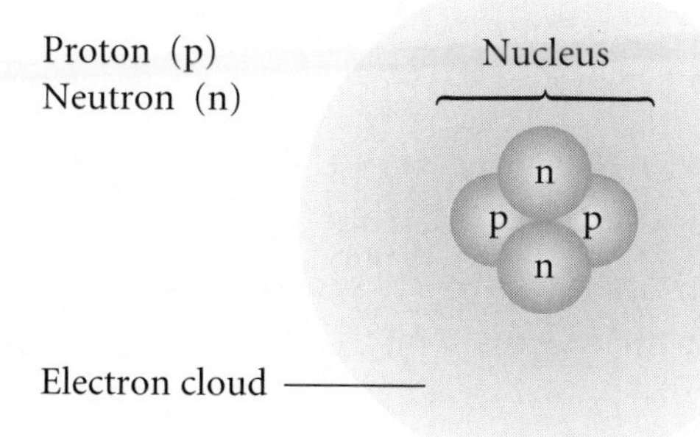

1

The protons (p) and the neutrons (n) are packed together in an inner core called the ____________. The outer part of the atom, which contains electrons, is called the ____________.

nucleus; electron cloud

a negative electrical charge, because the electron cloud consists of electrons

2

The electron cloud has a negative electrical charge. What type of charge would you expect the electron to have? ______________________________

____

a. repel
b. attract
c. repel
d. attract

3

The electron has a *negative electrical charge* and is symbolized by e^-. Remember that *like* electrical charges repel each other, and *unlike* charges attract.

Indicate whether the following pairs of charges would attract or repel each other.

a. ⊕ ⊕ ____________ b. ⊖ ⊕ ____________

c. ⊖ ⊖ ____________ d. ⊕ ⊖ ____________

positive

4

The nucleus attracts the negatively charged electrons. Therefore, the overall charge of the nucleus must be ____________ (negative/positive).

proton

5

The neutron was named for its electrical characteristics. It has no electrical charge; it is neutral. This means that the positive charge of the nucleus must be due to the *second type of particle* it contains. This second type of particle is the ____________.

6

So far, then, we have this picture of atomic structure:

a. An atom consists of an inner part, or ____________ that is made up of ____________ and ____________.

b. The electron has what type of charge? ____________

c. The proton has what type of charge? ____________

d. The neutron has a charge of ____________.

a. nucleus; protons; neutrons
b. negative
c. positive
d. zero (0)

7

The charge on the electron balances the charge on the proton. If the electron has a charge of –1, then the proton would have a charge of ____________ (−1, +1, ±1).

+1

8

An atom with one proton in its nucleus and one electron outside that nucleus would therefore have an overall charge of ____________ (+1, −1, 0).

0

9

Atoms are electrically neutral. This means that an atom will contain: (check one)

__ more protons than electrons
__ more electrons than protons
__ an equal number of protons and electrons

an equal number of protons and electrons

12

10

An atom with 12 protons in the nucleus would have how many electrons outside the nucleus? ____________

a. proton
b. electron

The atom with the simplest atomic structure is hydrogen. (For simplicity we shall merely indicate the electron(s) outside the nucleus, and omit the electron cloud.)

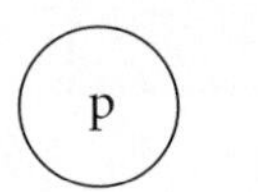

Hydrogen

a. The nucleus of the hydrogen atom consists of one ____________.

b. The outer part of the atom, the electron cloud, contains one ____________.

2; 2; 2

The helium atom is a little more complicated.

e⁻ n p p n e⁻

Helium

It contains: (how many?)

____________ neutrons
____________ protons
____________ electrons

ATOMIC NUMBER

There are over 100 known elements. Each element has two numbers associated with it—numbers that give certain facts about the structure of its atoms.

The first number is the atomic number. This is the number of protons in the nucleus of the atom.

13

Hydrogen, the simplest atom, contains only one proton, so the atomic number of hydrogen is ____________.

1

14

Uranium is the most complicated of the elements that occur naturally. A uranium atom contains 92 protons, 146 neutrons, and 92 electrons. The atomic number of uranium is ____________.

92

15

An atom of magnesium, atomic number 12, must have a nucleus containing ____________ protons.

If the nucleus contains 12 protons, there must be how many electrons? ____________

12; 12

16

Therefore, the atomic number of an element indicates the number of ____________ in the nucleus of the atom, and also the number of ____________ outside the nucleus.

protons; electrons

MASS NUMBER

The second number associated with each atom is the *mass number*. The mass number expresses the sum of the masses of the particles in the atom.

A proton has a mass of 1 atomic mass unit. An electron is considered to have zero mass, or a mass of 0.

1

17

A hydrogen atom has a mass of ____________. (If you don't know, see problem 11.)

a. 2
b. 0
c. 2
d. 1

18

The helium atom has a mass number of 4.

a. The 2 protons in the helium atom have a total of how many atomic mass units? ____________

b. The 2 electrons in the helium atom have a total of how many atomic mass units? ____________

c. Therefore, for the helium atom to have a mass number of 4, the 2 neutrons must contain how many atomic mass units? ____________

d. If 2 neutrons have a total of 2 atomic mass units, a neutron must have an atomic mass of ____________.

in the nucleus

19

Because the electrons, which have practically no mass, are located outside the nucleus, the entire mass of the atom can be considered to be located:

__ in its electron cloud __ in the nucleus

20

The *atomic number* indicates the number of protons (each with atomic mass 1) inside the nucleus of an atom. The *mass number* indicates the number of protons and neutrons (each with atomic mass 1) in the nucleus. Therefore, the number of neutrons can be determined by *subtracting* the atomic number from the mass number.

The sodium atom has a mass number of 23 and an atomic number of 11. The number of neutrons in the nucleus of the sodium atom is

____________.

12

21

The carbon atom has an atomic number of 6 and a mass number of 12. The carbon atom contains: (how many?)

____________ protons in its nucleus
____________ neutrons in its nucleus
____________ electrons outside its nucleus

6; 6; 6

22

The element phosphorus has an atomic number of 15 and a mass number of 31. Indicate on the blank lines in the diagram the number of protons, neutrons, and electrons.

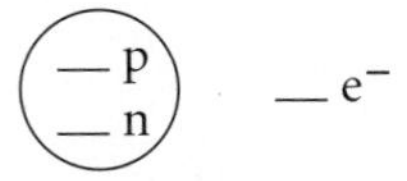

Phosphorus

15p
16n
$15e^-$

12p
12n
$12e^-$

23

Diagram the structure of the magnesium atom, atomic number 12 and mass number 24.

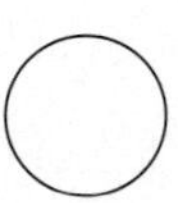

Magnesium

92; 238

24

The uranium atom has the atomic structure shown here.

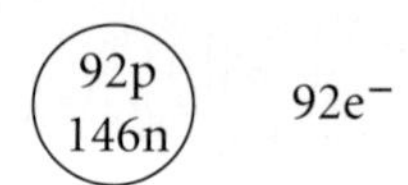

Uranium

Therefore, the uranium atom has an atomic number of __________ and a mass number of ____________.

ISOTOPES

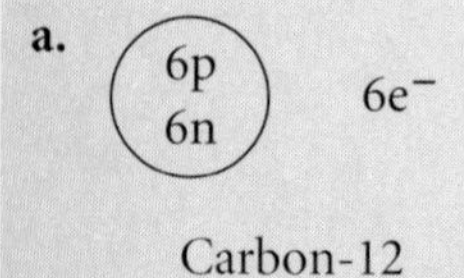

Carbon-12

25

a. Draw the structure of a carbon atom, atomic number 6 and mass number 12.

b. Draw the structure of a carbon atom, atomic number 6 and mass number 13.

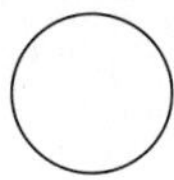

b. 6p 7n 6e$^-$

Carbon-13

26

Here are the structures you drew for the two carbon atoms:

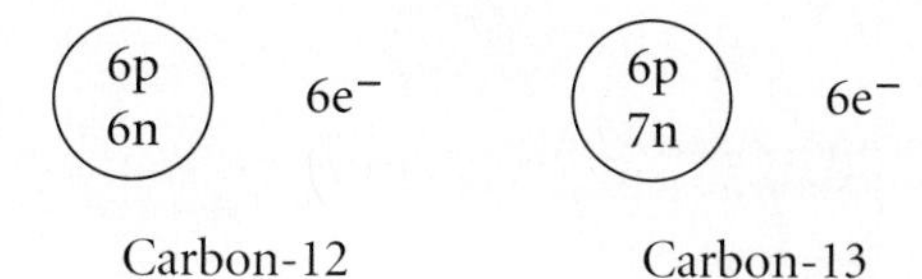

Carbon-12 Carbon-13

a. These atoms have ____________ (the same/different) atomic number(s).

b. These atoms have ____________ (the same/different) mass number(s).

Such atoms are called *isotopes.*

a. the same
b. different

27

Isotopes, then, may be defined as atoms that have:

__ the same atomic number and the same mass number
__ different atomic numbers
__ different mass numbers and the same atomic number

different mass numbers and the same atomic number

a.

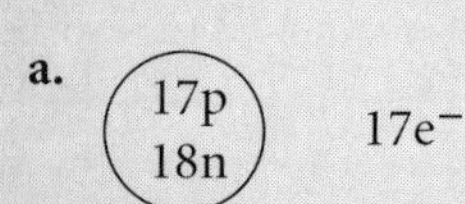

Chlorine-35

b.

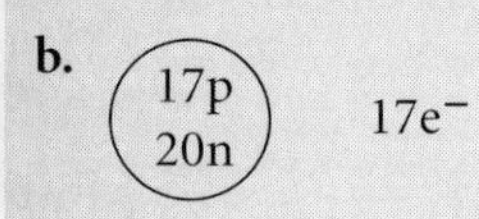

Chlorine-37

a.

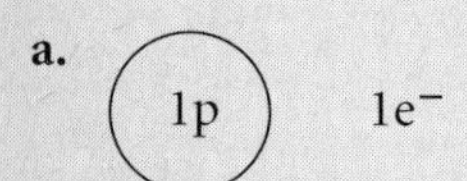

Hydrogen-1

b.

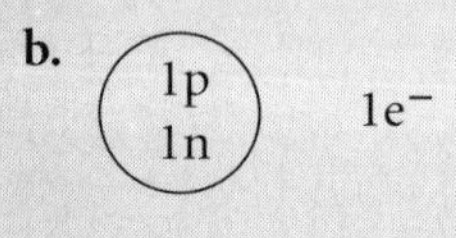

Hydrogen-2

c.

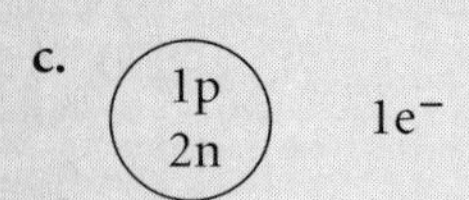

Hydrogen-3

a. same
b. different

28

Draw the two isotopes of chlorine, atomic number 17 and mass numbers 35 and 37.

a. chlorine-35 b. chlorine-37

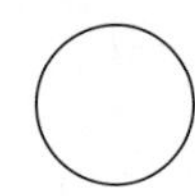
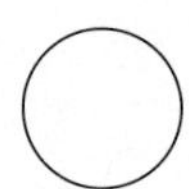

29

Draw the three isotopes of hydrogen, atomic number 1 and mass numbers 1, 2, and 3.

a. hydrogen-1 b. hydrogen-2 c. hydrogen-3

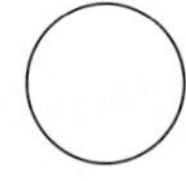
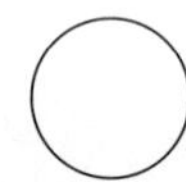
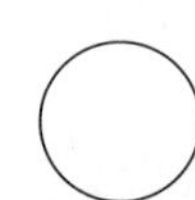

30

Radioisotopes, isotopes that give off radiation, are frequently used in medical applications. I-131 (radioactive iodine, also written as ^{131}I) is used in the diagnosis and treatment of thyroid conditions. How does I-131, radioactive iodine, compare with I-127, nonradioactive iodine, in

a. atomic number ____________

b. mass number ____________

ELECTRON ENERGY LEVELS

The electrons are located outside of the nucleus of the atom. These electrons make up the electron cloud, which may be subdivided into different energy levels. The first energy level is nearest the nucleus; then comes the second energy level, the third energy level, and so on.

Each energy level can hold a certain maximum number of electrons. This maximum number may be determined by using the formula $X = 2n^2$ (X is the maximum number of electrons in energy level number n).

31

Using the formula $X = 2n^2$, if $n = 1$, then $X =$ ____________. The energy level indicated by $n = 1$ is the first energy level. Therefore, the first energy level can hold a maximum of ____________ electrons.

2; 2

32

a. Can the first energy level hold fewer than two electrons? __________

b. Can the first energy level hold more than two electrons? __________

a. yes
b. no

33

For the second energy level, where $n = 2$, the maximum number of electrons is ____________.

8

34

a. The maximum number of electrons in the third energy level is ____________.

b. The maximum number of electrons in the fourth energy level is ____________.

a. 18
b. 32

$2e^-$ in the first energy level; $8e^-$ in the second energy level; $18e^-$ in the third energy level; $32e^-$ in the fourth energy level

35

Label the maximum number of electrons possible in each energy level in the diagram.

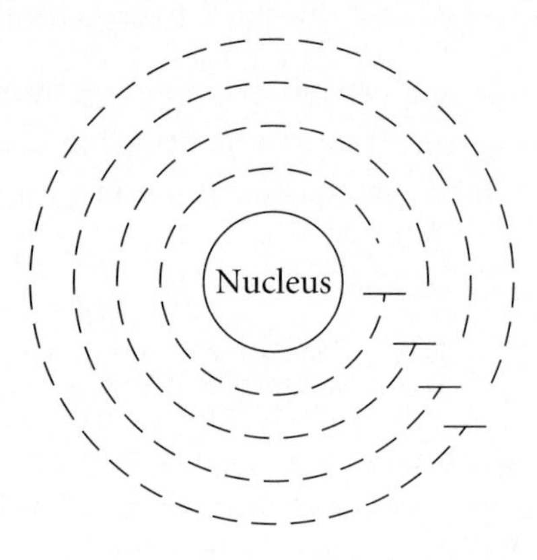

8

36

The following rules must be observed when considering the placement of electrons in the various energy levels. The first energy level must be filled with 2 electrons before electrons can go into the second energy level.

The second energy level must be filled with ____________ electrons before electrons can go into the third energy level.

For elements having more than three energy levels, the sequence of filling those levels is complex and can be found in a general chemistry text.

37

On the diagram, show the structure of the hydrogen atom, atomic number 1 and mass number 1.

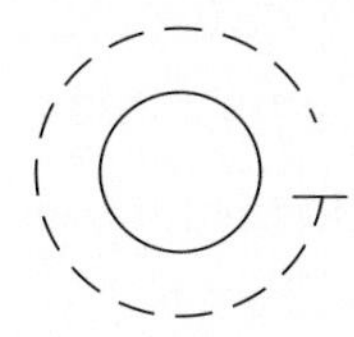

Hydrogen atom

a. The number of protons in the hydrogen atom is ____________.

b. The number of neutrons is ____________.

c. The number of electrons is ____________.

d. The 1 electron in the hydrogen atom must go into which energy level—1st, 2nd, or 3rd? ____________

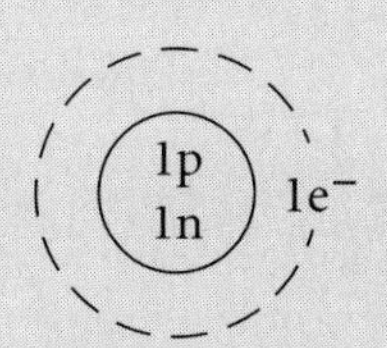

Hydrogen atom

a. 1
b. 0
c. 1
d. 1st

38

In the space provided, draw the structure of the helium atom, atomic number 2 and mass number 4.

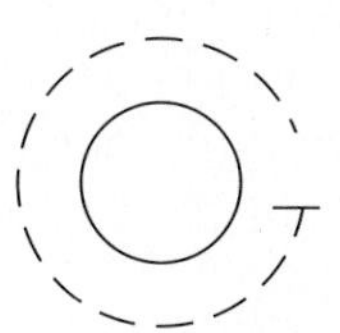

Helium atom

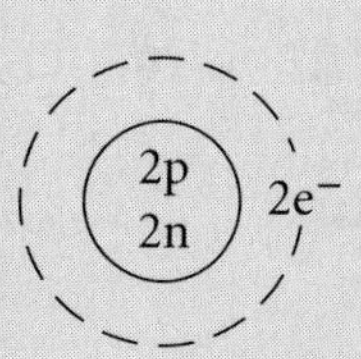

Helium atom

a. 3
b. 4
c. 3

39

Now let's look at the structure for the lithium atom, atomic number 3 and mass number 7.

a. The number of protons in the lithium atom is ____________.

b. The number of neutrons in the lithium atom is ____________.

c. The number of electrons in the lithium atom is ____________.

2

40

There are 3 electrons in the lithium atom. How many energy levels will the lithium atom have? ____________ (If you aren't sure, check problem 36.)

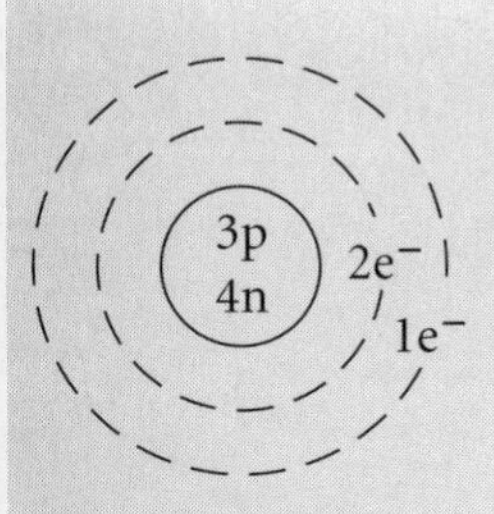

Lithium atom

41

Complete the diagram of the lithium atom.

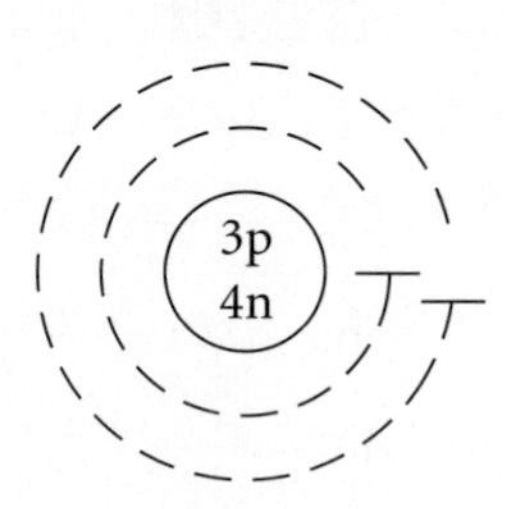

Lithium atom

42

The carbon atom has atomic number 6 and mass number 12.

a. The carbon atom contains ____________ protons.

b. The carbon atom contains ____________ neutrons.

c. The carbon atom contains ____________ electrons.

d. Complete the structure of the carbon atom.

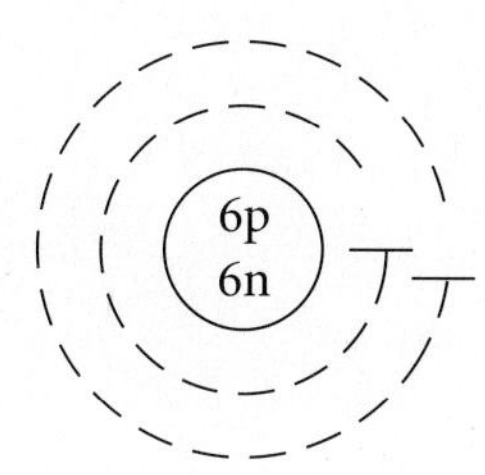

Carbon atom

a. 6
b. 6
c. 6

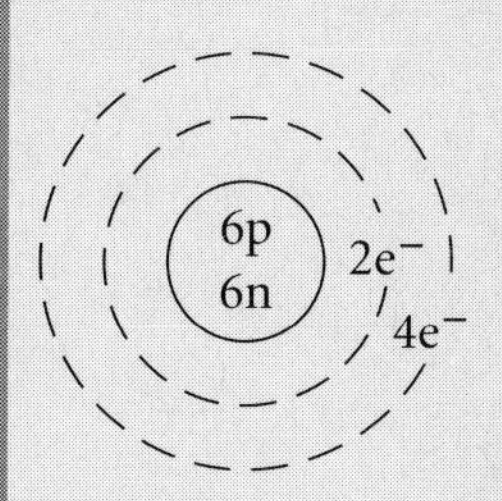

Carbon atom

43

a. In the element sodium, atomic number 11, there are how many electrons? ____________

b. How many energy levels will the sodium atom have? ____________

c. How many electrons will each energy level of the sodium atom have?

1st energy level __ 2nd energy level __ 3rd energy level __

d. Diagram the complete structure of the sodium atom, atomic number 11 and mass number 23.

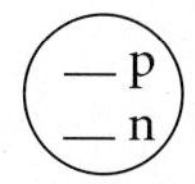

Sodium atom

a. 11
b. 3
c. 2; 8; 1

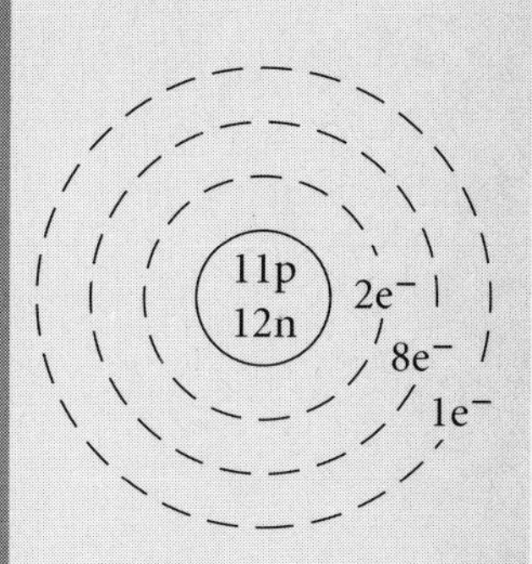

Sodium atom

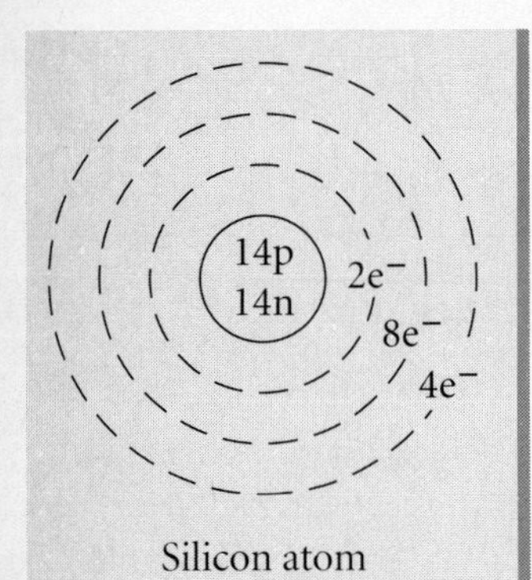

Silicon atom

44

Diagram the structure of the silicon atom, atomic number 14 and mass number 28.

Chemical Symbols 2

Elements are the building blocks of the universe. They cannot be broken down into simpler substances by ordinary chemical means. Each of the over 100 elements has its own name and its own properties. The symbol for an element is usually an abbreviation for its name. Frequently the abbreviation is merely the first letter of that name. The following table lists some of these elements.

Element	Symbol
Hydrogen	H
Carbon	C
Oxygen	O
Nitrogen	N
Phosphorus	P
Sulfur	S

45

What elements are present in water, H_2O? ____________

hydrogen and oxygen

46

What elements are present in ammonia, NH_3? ____________

nitrogen and hydrogen

carbon, hydrogen, and oxygen

47

What elements are present in glucose, $C_6H_{12}O_6$? ____________

carbon, hydrogen, oxygen, nitrogen, sulfur

48

Biotin, one of the B-vitamins, $C_{10}H_{16}O_3N_2S$, contains which elements?

____

carbon, hydrogen, oxygen, nitrogen, phosphorus

49

Adenosine triphosphate, ATP, the body's primary energy compound, has the formula $C_{11}H_{18}O_{13}N_5P_3$. What elements does it contain? __________

Ca; Br; Si; Ba

50

When the names of more than one element begin with the same letter, frequently the second letter of the name is added to the symbol. Note that only the first letter of the symbol is capitalized.

Here are some of the elements whose symbols are the first two letters of their names. Write the symbols for each.

Element	Symbol
Calcium	____________
Bromine	____________
Silicon	____________
Barium	____________

calcium and bromine

51

$CaBr_2$ contains which two elements? ____________

52

BaO_2 contains which two elements? ____________

barium and oxygen

53

CaC_2O_4 contains which three elements? ________________

calcium, carbon, and oxygen

There are several elements whose symbols are not derived from the first letter or first two letters of their English names. Some of those elements are listed in the following table.

Element	Symbol
Chlorine	Cl
Sodium	Na (from the Latin *natrium*)
Magnesium	Mg
Potassium	K (from the Latin *kalium*)
Zinc	Zn
Iron	Fe (from the Latin *ferrum*)

54

Write the name of the element represented by each of the following symbols:

H ____________	Na ____________	Si ____________
N ____________	K ____________	Br ____________
O ____________	C ____________	Zn ____________
Ca ____________	S ____________	Fe ____________

***H*ydrogen; *N*itrogen; *O*xygen; *Ca*lcium; Sodium; Potassium; *C*arbon; *S*ulfur; *Si*licon; *Br*omine; *Z*i*n*c; Iron**

a. potassium and bromine
b. carbon and chlorine
c. sodium, oxygen, and hydrogen
d. magnesium and chlorine
e. sodium, nitrogen, and oxygen
f. potassium, iron, carbon, and nitrogen

carbon, hydrogen, oxygen, nitrogen, and magnesium

55

What elements are present in each of the following?

a. KBr ____________ b. CCl_4 ____________

c. NaOH ____________ d. $MgCl_2$ ____________

e. $NaNO_3$ ____________ f. $K_3FeC_6N_6$ ____________

56

Chlorophyll has the formula $C_{55}H_{68}O_5N_4Mg$. It contains which elements? ______________________________

Atoms and Molecules 3

57

One atom of hydrogen is written as H. The number 1 is understood. Two atoms of hydrogen are written as 2H. Three atoms of carbon are written as ____________.

3C

58

Atoms combine to form *molecules*. A molecule contains two or more atoms of the same or different elements. The molecule O_2 contains how many atoms of oxygen? ____________

2

59

Each molecule of glucose, $C_6H_{12}O_6$, contains:

____________ atoms of carbon
____________ atoms of hydrogen
____________ atoms of oxygen

6; 12; 6

60

A water molecule is composed of two hydrogen atoms and one oxygen atom. Write the formula for a water molecule. ____________

H_2O

$C_{12}H_{22}O_{11}$

61

A molecule of sucrose contains 12 atoms of carbon, 22 atoms of hydrogen, and 11 atoms of oxygen. The formula for sucrose is ____________.

$C_{63}H_{90}O_{14}N_{14}PCo$

62

One molecule of vitamin B_{12} contains:

63 atoms of carbon
90 atoms of hydrogen
14 atoms of oxygen
14 atoms of nitrogen
1 atom of phosphorus
1 atom of cobalt (Co)

The formula for vitamin B_{12} is ____________.

a. 6
b. $8H_2O$

63

When you want to represent more than one molecule of a compound, the number of molecules precedes the symbols.

a. Thus, $6CO_2$ indicates how many molecules of carbon dioxide? ____________

b. How would eight molecules of water be represented? ____________

molecule; hydrogen; 2

64

The formula H_2 indicates one ____________ of ____________, composed of how many atoms? ____________

65

The formula $2H_2$ indicates ____________ of hydrogen.

two molecules

66

What do the following formulas represent?

$3O$ ____________ $4N_2$ ____________ $7H_2O$ ____________

3 atoms of oxygen;
4 molecules of nitrogen;
7 molecules of water

Ionization 4

IONS

67

Atoms contain equal numbers of protons and electrons. Thus, atoms are:

__ electropositive   __ electronegative   __ electrically neutral

electrically neutral

68

The electron is a ____________ (negatively/positively) charged particle. So, if an atom loses an electron, it will then have an overall ____________ (negative/positive) charge.

negatively; positive

69

If an atom were to *gain* an electron, it would then have an overall ____________ charge.

negative

70

An *ion* is an atom that has acquired an electrical charge by either losing or gaining an ____________.

electron

positively

71

Note that electrons, which are located outside the nucleus, are lost or gained in the formation of ions. The nucleus, which contains protons and neutrons, does not take part in the formation of ions.

If a hydrogen atom loses its electron, it becomes a hydrogen ion. Because it loses an electron, it is a ____________ (negatively/positively) charged ion.

a. ion
b. negatively
c. positively

72

a. Any atom that has lost or gained an electron is an ____________.

b. If an atom gains an electron, it becomes a ____________ charged ion.

c. If an atom loses an electron, it becomes a ____________ charged ion.

valence shell

73

Some elements reach their most stable state when they have 2 electrons in their outer (their only) energy level, which is called the *valence shell.*

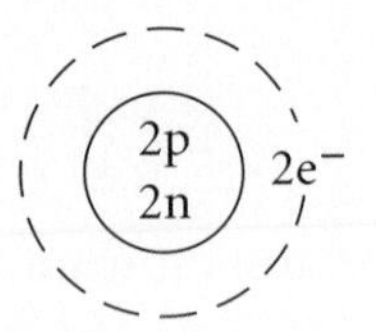

Helium atom

This atom of helium, for example, contains 2 electrons, which fill the first energy level. So, the first energy level in helium is called the ____________.

74

Helium (He) has a completed valence shell, one that can hold no more electrons. Thus, helium is a very ____________ (stable/unstable) atom. Such elements are usually unreactive.

stable

75

Most atoms reach their most stable state when they have 8 electrons in their outer energy level. This is called the octet rule.

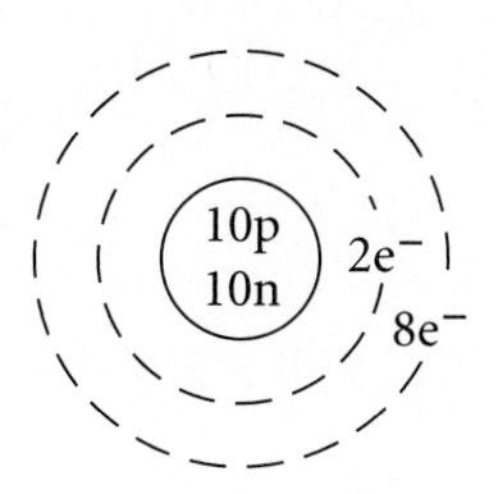

Neon atom

This atom of neon (Ne), for example, contains 10 electrons. Two of the electrons fill the first energy level, and the remaining 8 electrons are in the second energy level. So the second energy level of neon would be called its ____________.

valence shell

a. valence shell; third; 1
b. Cl

76

Look at this atom of chlorine.

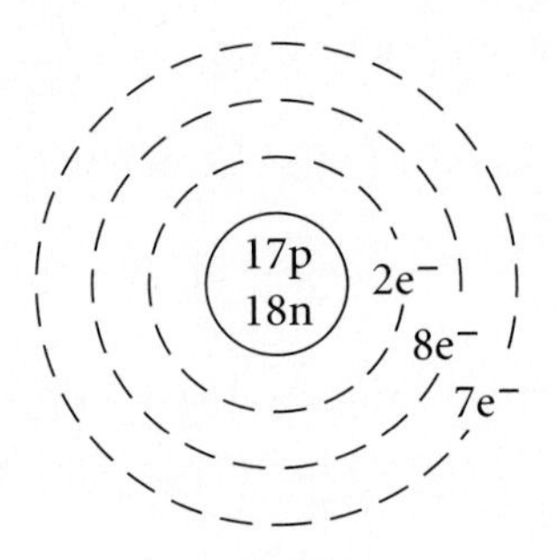

Chlorine atom

a. In chlorine, the outer energy level, or ____________, is the ____________ energy level. It contains 7 electrons. Thus, to attain a stable state, the chlorine atom needs to gain how many electrons? ____________

b. What is the symbol for chlorine? ____________

a. ion
b. negative

77

Suppose that an atom of chlorine gains the 1 electron it needs to become stable.

a. Because it has gained an electron, it is now a charged atom, called a(n) ____________.

b. Would it have a positive or negative charge? ____________

Cl^-

78

Which of the following symbols stands for the chloride ion (a chlorine atom that has gained 1 electron)?

__ Cl __ Cl^+ __ Cl^-

79

If a chlorine atom is to gain an electron, some other atom must lose the electron.

a. That atom could be sodium, the symbol for which is ____________.

b. The sodium atom has how many electrons in its valence shell?

____

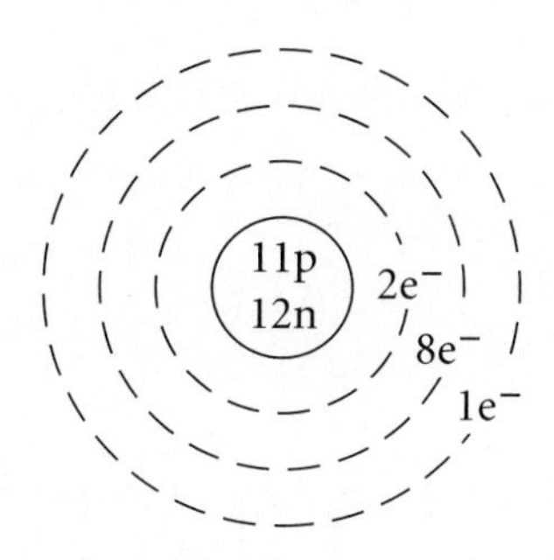

Sodium atom

c. To become stable it will have to ____________ (gain/lose) one electron.

a. Na
b. 1
c. lose

80

When the sodium atom loses 1 electron, it becomes a ____________ (positively/negatively) charged ____________.

positively; ion

81

Which symbol should we use for the sodium ion?

__ Na __ Na^+ __ Na^-

Na^+

a. +1
b. −1
c. +2

82

Positive and negative charges are indicated by superscript plus (positive) or minus (negative) signs, respectively, with the number 1 being understood. Charges greater than 1 are written as superscripts, with the number preceding the plus or minus sign.

a. Na^+ indicates a sodium ion with a charge of ____________.

b. Cl^- indicates a chloride ion with a charge of ____________.

c. Mg^{2+} indicates a magnesium ion with a charge of ____________.

a. Al^{3+}
b. S^{2-}

83

a. How would an aluminum ion with a charge of +3 be indicated?

____

b. How would a sulfide ion with a charge of −2 be indicated?

____

loses; positive; 1;
gains; negative; 1

84

$$Na + Cl \longrightarrow Na^+ + Cl^-$$

Reaction of sodium and chlorine

When a sodium atom reacts with a chlorine atom, the sodium atom ____________ (gains/loses) 1 electron to form a sodium ion with a ____________ charge of ____________ (how many?).

At the same time, the chlorine atom ____________ (gains/loses) 1 electron to form an ion with a ____________ charge of ____________ (how many?).

85

Mg; 2; lose; 2; Mg^{2+}

Look at this magnesium atom. Its symbol is ____________.

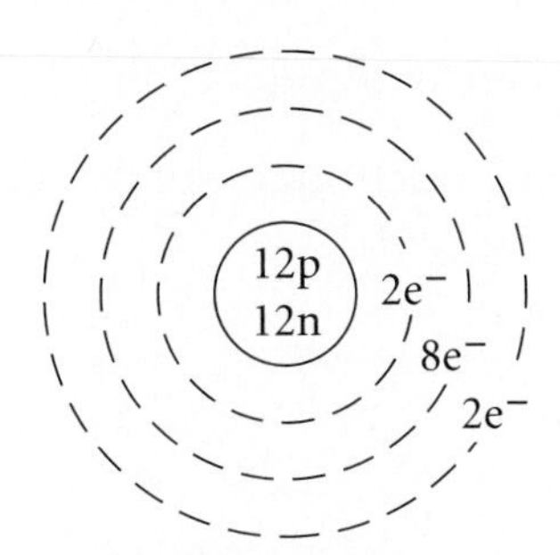

Magnesium atom

It has ____________ electrons in its valence shell. So, to gain stability, it will have to ____________ (gain/lose) ____________ (how many?) electrons.

The symbol for the magnesium ion is ____________ (Mg^{2-}, Mg, Mg^{2+}).

86

−1; +2

The reaction between magnesium and chlorine may be written as:

$$Mg + Cl_2 \longrightarrow Mg^{2+} + 2Cl^-$$

This reaction shows that 1 chlorine molecule reacts with 1 magnesium atom to form 2 chloride ions, each with a charge of ____________, and 1 magnesium ion with a charge of ____________.

a. 3
b. 6
c. −2
d. S^{2-}

87

a. The sulfur atom, atomic number 16, has how many energy levels? ____________

b. How many electrons are in its valence shell? ____________

c. If a sulfur atom gains 2 electrons to fill its valence shell, it will have a charge of ____________.

d. The symbol for the sulfide ion is ____________.

88

How can we tell whether an atom will lose or gain electrons to reach a stable structure of 8 electrons in its valence shell? Look over the following diagrams.

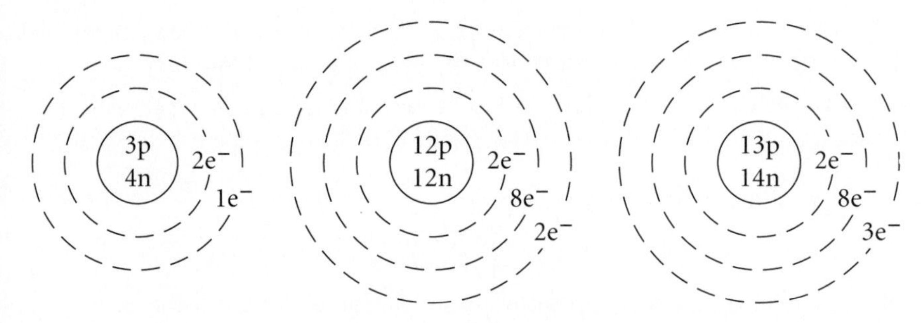

Form positive ions

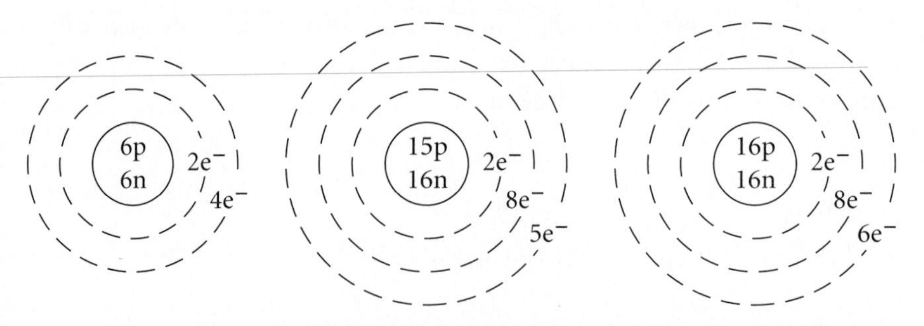

Do not form positive ions

In general, an atom with ____________ (how many?) electrons in its valence shell tends to lose electrons and form a ____________ (positively/negatively) charged ion.

1, 2, or 3 (or equivalent answer); positively

89

Sodium has atomic number 11. How many electrons does it have in its valence shell? ____________

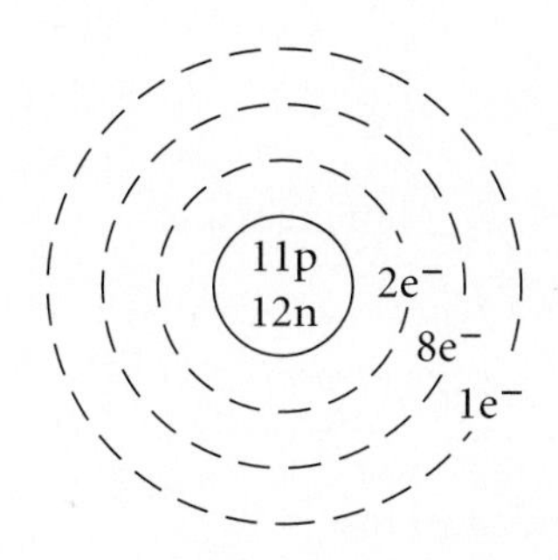

Sodium atom

The sodium ion, therefore, is formed by the ____________ (loss/gain) of 1 electron.

The symbol for the sodium ion is ____________.

1; loss; Na^+

90

Here is the hydrogen atom, atomic number 1 and mass number 1.

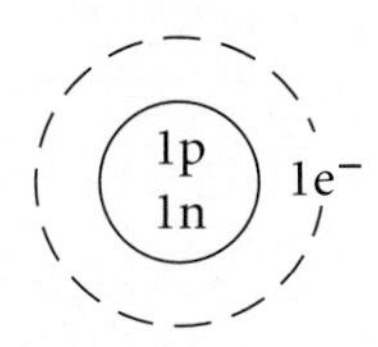

Hydrogen atom

Hydrogen usually loses its 1 electron to form a hydrogen ion with a charge of ____________.

The hydrogen ion can be represented as ____________.

+1; H^+

3; loss; Al^{3+}

91

Here is an aluminum atom:

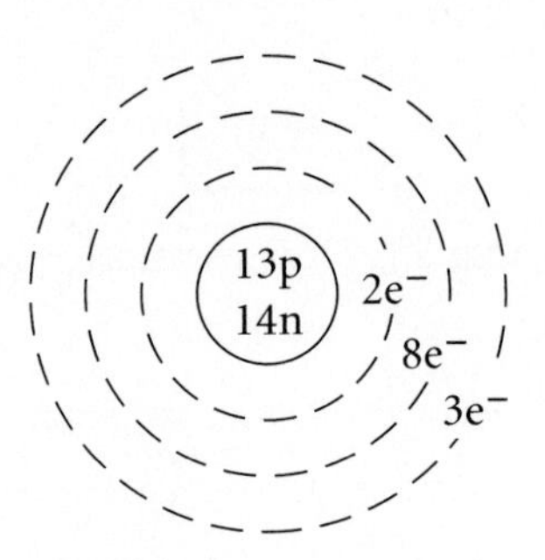

Aluminum atom

Its valence shell contains _________ electrons. The aluminum ion is formed by the ____________ (loss/gain) of 3 electrons.

The symbol for the aluminum ion is ____________.

Mg^{2+}

92

Magnesium, atomic number 12, forms a magnesium ion that may be represented as ____________.

93

Look at the following diagrams carefully.

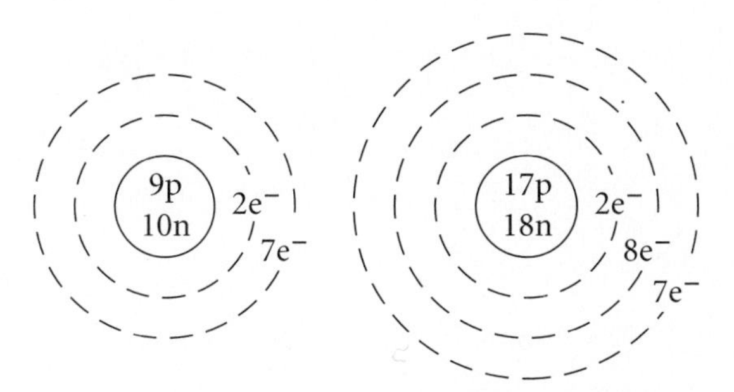

Form negative ions

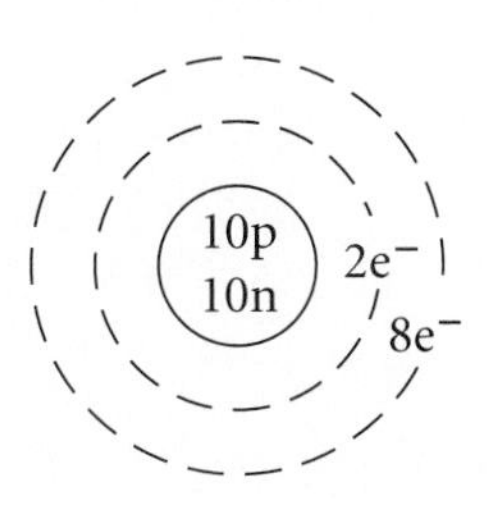

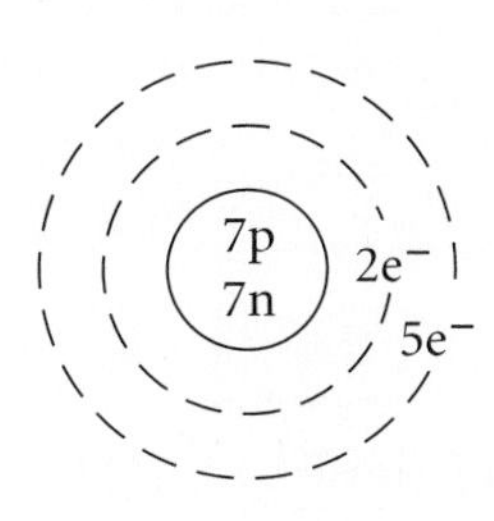

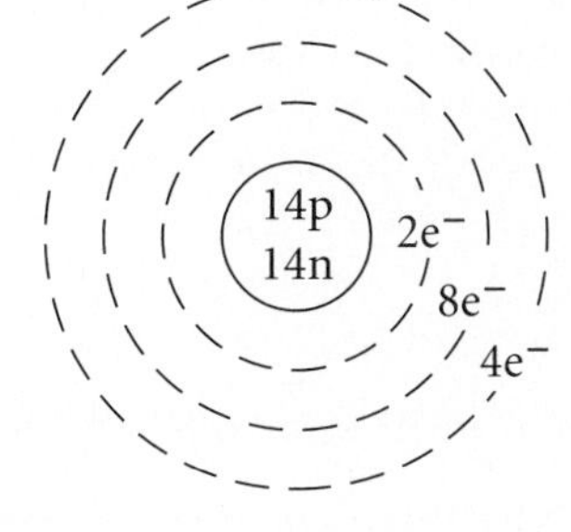

Do not form negative ions

In general, atoms that have ____________ (how many?) electrons in their valence shells form negatively charged ions.

6 or 7

94

When an atom gains electrons to fill its valence shell, it forms _________ charged ions.

negatively

9; 7; gain; 1; F^-

95

Complete the following for the fluorine (F) atom shown here:

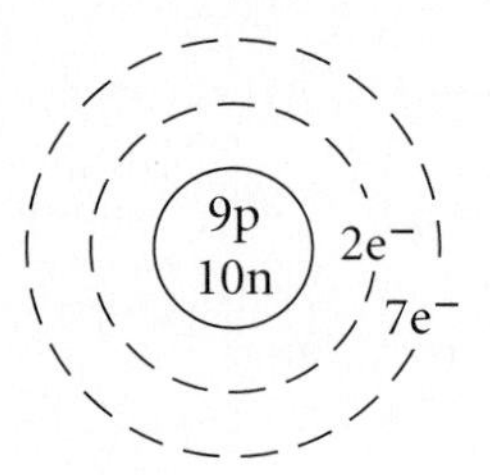

Fluorine atom

Atomic number: ____________
Number of electrons in valence shell: ____________
Ion formed by ____________ (gain/loss) of ____________ electron(s)
Symbol for ion: ____________

S; 16; 32; S^{2-}

96

Complete the following for the sulfur atom shown here:

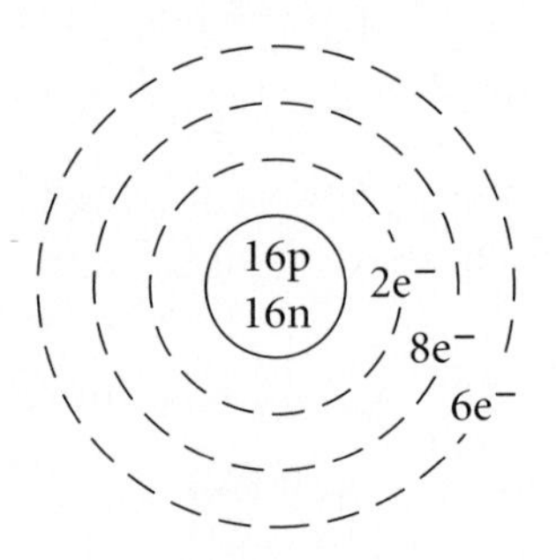

Sulfur atom

Symbol for sulfur: ____________
Atomic number: ____________
Mass number: ____________
Symbol for ion: ____________

97

Atoms with 4 or 5 electrons in their valence shells usually do not form ions. These we shall discuss later.

Atoms with 8 electrons in their valence shells are stable. Because they do not have to gain or lose electrons, you would expect these stable atoms to be:

__ reactive __ inert (unreactive)

inert (unreactive)

98

The sodium atom, atomic number 11, has how many electrons in its valence shell? ____________

The sodium atom will tend to lose the 1 electron in its valence shell to form an ion with a charge of ____________.

1; +1

99

The chlorine atom, which has the symbol ____________, has atomic number 17. It has how many electrons in its valence shell? ____________

The chlorine atom will tend to gain ____________ (how many?) electron(s) to form an ion with a charge of ______________.

Cl; 7; 1; −1

IONIC BONDS

gains

100

Consider the reaction between sodium and chlorine:

$$Na + Cl \longrightarrow Na^+ + Cl^-$$

The sodium ion and the chloride ion are oppositely charged. These ions are held together by the attraction of their opposite charges. We say that there is an *ionic bond* between the sodium ion and the chloride ion.

An ionic bond is produced whenever one atom loses an electron or electrons and another atom ____________ an electron or electrons.

ionic bond

101

In the reaction $Zn + S \longrightarrow Zn^{2+} + S^{2-}$, the zinc ion and the sulfur ion are held together by an ____________.

ion; negatively; ion

102

When a sodium atom combines with a chlorine atom according to the equation $Na + Cl \longrightarrow Na^+ + Cl^-$, a compound containing a positively charged sodium ____________ and a ____________ charged chloride ____________ is formed. This compound (sodium chloride) is usually written as NaCl, with the charges being understood and not written.

chloride; Cl^-

103

When the compound NaCl is placed in water, the ionic bond holding the sodium ion and the chloride ion together is weakened, so that these ions are free to move throughout the solution.

Therefore, NaCl in water produces a *solution* containing sodium ions (Na^+) and ____________ ions (____________).

ACIDS, BASES, AND SALTS

104

a. When the compound HCl is placed in water, it produces a solution containing ____________ ions (____________) and ____________ ions (____________).

b. When the compound H_2SO_4 is placed in water, it produces a solution containing ____________ ions and HSO_4^- ions.

a. hydrogen, H^+, chloride, Cl^-
b. hydrogen (H^+)

105

Any substance that yields hydrogen ions (H^+) in solution is called an *acid.*

$$\begin{aligned} HCl &\longrightarrow H^+ + Cl^- \\ H_2SO_4 &\longrightarrow H^+ + HSO_4^- \\ HNO_3 &\longrightarrow H^+ + NO_3^- \end{aligned}$$

HNO_3 is an *acid* because it yields ____________ in solution.

hydrogen ions or H^+

106

A hydrogen ion is also a *proton*. Therefore, acids yield hydrogen ions or ____________ in solution.

protons

acids

107

Note that hydrogen ions associate with water molecules to form a "hydronium" ion $(H_3O)^+$. However, we will use the term *hydrogen ion* (H^+) in this book.

Substances that yield hydrogen ions or protons in solution are called ____________. Substances that accept (react with) hydrogen ions or protons are called *bases*.

bases

108

When sodium hydroxide, NaOH, is placed in water, sodium ions and hydroxide ions are present:

$$NaOH \longrightarrow Na^+ + OH^-$$

The hydroxide ions (OH^-) react with hydrogen ions to form water:

$$OH^- + H^+ \longrightarrow H_2O$$

Because hydroxide ions accept (react with) hydrogen ions, OH^- ions are __________.

yes

109

Potassium hydroxide, KOH, also yields OH^- ions in water. Would KOH be a base? ____________

bases; they accept (react with) hydrogen ions

110

Bicarbonate ions, HCO_3^-, are important ions in body fluids. They react as follows:

$$HCO_3^- + H^+ \longrightarrow H_2CO_3$$

Are bicarbonate ions acids or bases? ____________

Why? ____

111

Ammonia is a waste product of the body's metabolism of protein. Ammonia reacts as follows:

$$\underset{\text{Ammonia}}{NH_3} + H^+ \longrightarrow \underset{\text{Ammonium ion}}{NH_4^+}$$

Is ammonia an acid or a base? ____________

base

112

Acids ____________ hydrogen ions or protons.

Bases ____________ hydrogen ions or protons.

yield; accept (react with)

113

When the compound NaCl is placed in water, it yields ____________ ions and ____________ ions.

sodium (or Na^+); chloride (or Cl^-)

114

The solution of NaCl in water would not be considered an acid because:

__ it yields OH^- ions
__ it doesn't break up into ions
__ it yields no hydrogen ions

it yields no hydrogen ions

a. salt
b. acid
c. salt
d. acid
e. base
f. acid
g. base

115

A compound that yields ions other than hydrogen ions (H^+) or hydroxide ions (OH^-) is called a *salt*. Are the following underlined substances acids, bases, or salts?

a. $\underline{Na_2CO_3} \rightarrow 2Na^+ + CO_3^{2-}$ ____________

b. $\underline{H_2SO_4} \rightarrow H^+ + HSO_4^-$ ____________

c. $\underline{MgCl_2} \rightarrow Mg^{2+} + 2Cl^-$ ____________

d. $\underline{H_2PO_4^-} \rightarrow H^+ + HPO_4^{2-}$ ____________

e. $\underline{NaOH} + H^+ \rightarrow Na^+ + H_2O$ ____________

f. $\underline{H_3PO_4} \rightarrow H^+ + H_2PO_4^-$ ____________

g. $\underline{Ca(OH)_2} + 2H^+ \rightarrow Ca^{2+} + 2H_2O$ ____________

ELECTROLYTES AND NONELECTROLYTES

yes

116

Acids, bases, and salts are called *electrolytes*. Solutions of electrolytes conduct electricity because of the presence of ions:

$$\underset{\text{Hydrochloric acid}}{HCl} \rightarrow H^+ + Cl^-$$

Would a solution of hydrochloric acid be an electrolyte? __________

117

Here's what happens when magnesium chloride is put in solution:

$$MgCl_2 \longrightarrow Mg^{2+} + 2Cl^-$$

a. Does the solution of $MgCl_2$ contain ions? ____________

b. Is $MgCl_2$ an acid, a base, or a salt? ____________

c. Is a solution of $MgCl_2$ an electrolyte? ____________

a. yes
b. salt
c. yes

118

When sucrose, $C_{12}H_{22}O_{11}$, is placed in water, no ions are produced.

a. Would a solution of sucrose conduct electricity? ____________

b. Would a solution of sucrose be an electrolyte or a nonelectrolyte?

____

a. no
b. nonelectrolyte

119

Which of the following solutions would be electrolytes?____________

a. $MgSO_4 \longrightarrow Mg^{2+} + SO_4^{\ 2-}$

b. alcohol $\longrightarrow$ no ions

c. $KOH \longrightarrow K^+ + OH^-$

d. $C_6H_{12}O_6$ (glucose) $\longrightarrow$ no ions

e. $HNO_3 \longrightarrow H^+ + NO_3^{\ -}$

a, c, e

ANIONS AND CATIONS

a. −
b. +

120

Consider the following electrolytic system, a battery whose two electrodes are immersed in a solution containing the indicated ions:

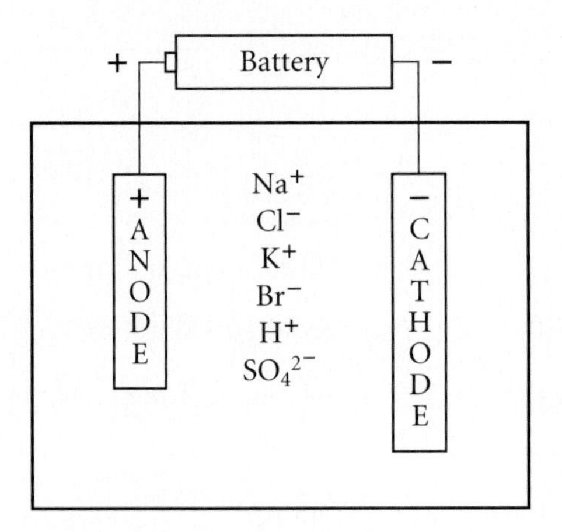

a. The anode will attract ions of what charge? ____________

b. The cathode will attract ions of what charge? ____________

a. −
b. +

121

a. Ions attracted towards an anode are called *anions*. Anions have a ____________ charge.

b. Ions attracted towards a cathode are called *cations*. Cations have a ____________ charge.

Na^+, K^+, and H^+

122

Which are the cations in the solution indicated in problem 120?

____

123

Which are the anions in that same solution? ____________

Cl^-, Br^-, and SO_4^{2-}

124

a. The principal cation inside animal cells (the principal *intra*cellular cation), the potassium ion, has a ____________ (positive/negative) charge.

b. The principal cation outside animal cells (the principal *extra*cellular cation), the sodium ion, has a ____________ (positive/negative) charge.

c. The iodide ion, I^-, is necessary for the proper functioning of the thyroid gland. The iodide ion is a(n) ____________ (anion/cation).

a. positive
b. positive
c. anion

pH

Mathematically, $pH = -\log H^+$. A log, or logarithm, is an exponent. The log (exponent) of 10^{-3} is -3.

125

The log of 10^{-7} is ____________.

−7

126

If $H^+ = 10^{-4}$, then $pH = -\log H^+ = -(-4) =$ ____________.

4

a. 12
b. 1

127

That is, pH is the negative log of the H^+.

a. If $H^+ = 10^{-12}$, pH = ____________.

b. If $H^+ = 10^{-1}$, pH = ____________.

a. acid
b. neutral
c. acid

128

The acid or basic strength of a solution may be expressed in terms of a number called the *pH* of that solution. The pH scale expresses the concentration of hydrogen ions (and hydroxide ions) in solution.

The pH range is from 0 to 14, with a pH of 7 indicating a neutral solution. A pH below 7 indicates an *acid* solution.

a. A pH of 3 indicates a(n) ____________ solution.

b. A pH of 7 indicates a(n) ____________ solution.

c. A pH of 6 indicates a(n) ____________ solution.

a. 7
b. neutral

129

Water is very slightly ionized: $H_2O \longrightarrow H^+ + OH^-$. The hydrogen ion concentration (H^+) in water is 10^{-7}.

a. What is the pH of water? ____________

b. Is water acid, basic, or neutral? ____________

Although all pH values below 7 indicate acid solutions, there is a definite progression of acid strengths according to pH values.

A pH between 5 and 7 indicates a weak acid solution, between 2 and 5 indicates a moderately strong acid solution, and between 0 and 2 indicates a strong acid solution, as shown in the following chart:

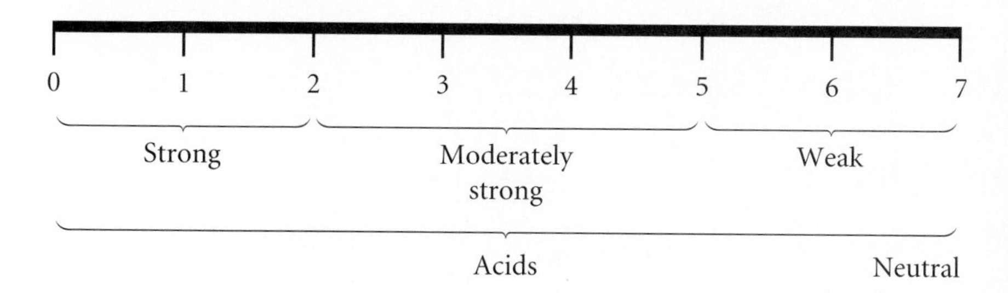

130

a. Which of the following pH values indicates a solution containing a strong acid?

__ 4 __ 7 __ 1 __ 6

b. Which of the following pH values indicates a solution containing a moderately strong acid?

__ 4 __ 7 __ 1 __ 6

c. A pH of 0 indicates what strength acid solution? ____________

d. A pH of 3 indicates what strength acid solution? ____________

a. 1
b. 4
c. strong
d. moderately strong

131

pH values may be indicated as decimal values, as well as whole numbers. Thus, a pH of 2.56 indicates a solution whose pH lies between 2 and 3. Thus, this is a:

__ strong acid solution
__ weak acid solution
__ moderately strong acid solution

moderately strong acid solution

6.27

132

Among the following pH values, which solution contains a weak acid?

__ 1.72 __ 3.75 __ 2.00 __ 6.27

1.72

133

Among the following pH values, which solution contains a strong acid?

__ 1.72 __ 3.75 __ 7.00 __ 5.00 __ 6.38

7.00

134

Of solutions with the following pH values, which one is neutral?

__ 2.70 __ 4.65 __ 5.00 __ 7.00

A solution whose pH is above 7 is called a *basic* solution.

A solution whose pH is between 7 and 9 is called a weak basic solution, between 9 and 12 a moderately strong basic solution, and between 12 and 14 a strong basic solution, as shown in the following chart:

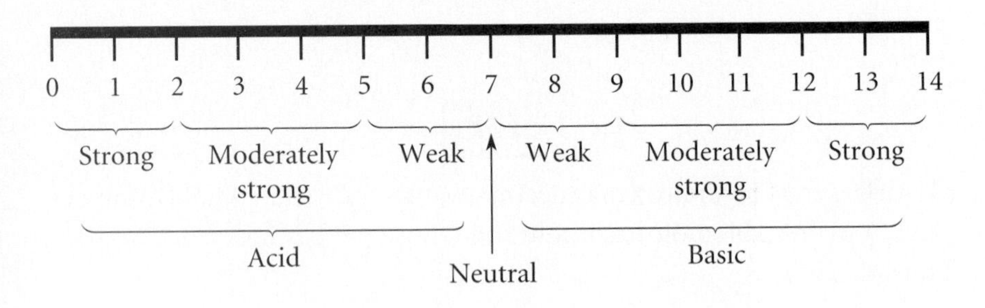

135

Of the following pH values, which indicates a strong basic solution?

__ 3.5 __ 7.0 __ 9.5 __ 13.9 __ 6.4

13.9

136

Of the following pH values, which indicates a weak basic solution?

__ 2.7 __ 8.3 __ 4.2 __ 10.8 __ 14.0

8.3

137

Which of the following pH values indicates a moderately strong basic solution?

__ 2.2 __ 4.7 __ 11.1 __ 13.7

11.1

138

The pH of saliva lies between 5.5 and 6.9, indicating a ____________ solution.

weak acid

139

Bile has a pH range of 7.8 to 8.6, so it is a ____________ solution.

weak basic

weak acid

140

Urine has a pH range of 5.5 to 6.9, so it is a ____________ solution.

strong acid

141

The gastric juices have a pH range of 1.6 to 1.8, so they make up a ____________ solution.

weak basic

142

Blood has a pH range of 7.35 to 7.45, so it is a ____________ solution.

weak basic

143

The pancreatic juices have a pH range of 7.5 to 8.0, so they make up a ____________ solution.

a. 10
b. weak

144

A difference of 1 in pH is equivalent to a 10-fold difference in acid or base strength. That is, an acid of pH 2 is 10 times as strong as an acid of pH 3. Likewise, a base of pH 10 is 10 times as strong as a base of pH 9.

a. A solution of pH 4.73 is ____________ times as strong as one of pH 5.73.

b. A solution of pH 9.81 is 10 times as ____________ (strong/weak) as one of pH 10.81.

145

Thus a small difference in pH can represent a ____________ (considerable/negligible) difference in acid or base strength.

considerable

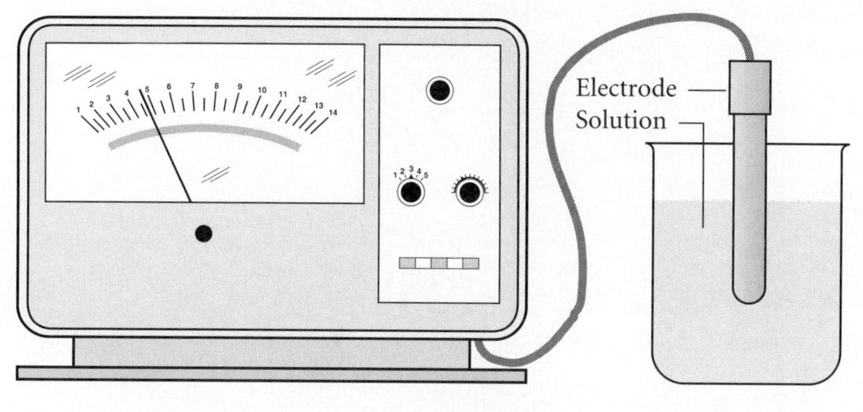

a pH meter

BUFFERS

146

Recall that acids yield hydrogen ions (H^+) or protons in solution. Both hydrochloric acid (HCl) and acetic acid (CH_3COOH), are acids, as indicated by the following reactions:

$$HCl \longrightarrow H^+ + Cl^-$$
$$CH_3COOH \rightleftharpoons CH_3COO^- + H^+$$

However, the HCl is 100% ionized, as shown by the one-way arrow, whereas CH_3COOH is less than 1% ionized, as shown by the double (equilibrium) arrows. That is, relatively speaking, HCl yields many H^+ ions in solution, whereas CH_3COOH yields few H^+ ions.

a. few
b. strong
c. weak

Hydrochloric acid, HCl, is called a *strong acid.* It yields relatively many H^+ ions in solution.

a. Acetic acid, CH_3COOH, is called a *weak acid.* It yields relatively ____________ H^+ ions in solution.

b. Sulfuric acid, H_2SO_4, yields relatively many H^+ ions in solution. It is a ____________ (strong/weak) acid.

c. Citric acid, which yields relatively few H^+ ions in solution, is a ____________ (strong/weak) acid.

a. many
b. few

147

A base accepts or reacts with protons (hydrogen ions).

a. A strong base reacts with relatively ____________ (many/few) H^+ ions.

b. A weak base reacts with relatively ____________ (many/few) H^+ ions.

a. strong
b. weak

148

a. NaOH, sodium hydroxide, reacts with relatively many H^+ ions. It is a ____________ (strong/weak) base.

b. Sodium bicarbonate, $NaHCO_3$, reacts with relatively few H^+ ions. It is a ____________ (strong/weak) base.

149

a. When a strong acid is added to water, the pH should go __________ (up/down).

b. When a strong base is added to water, the pH should go __________ (up/down).

The addition of a weak acid or a weak base to water should have very little effect on the pH.

a. down
b. up

150

A *buffer solution* is one that maintains a constant pH upon the addition of small amounts of either acid or base. A buffer consists of a weak acid and a weak base. A buffer solution can "sponge up" excess H^+ if added to a solution, or it can release H^+ if the H^+ concentration drops. One such buffer in the body is the $H_2CO_3/NaHCO_3$ system.

Suppose that a strong acid such as HCl is added to the buffer. The reaction is:

$$\underset{\text{Strong acid}}{HCl} + \underset{\text{Weak base}}{NaHCO_3} \longrightarrow \underset{\text{Weak acid}}{H_2CO_3} + \underset{\text{Salt}}{NaCl}$$

Thus the strong acid has been changed to a weak acid. Should the pH change? __________

no

151

If a strong base such as sodium hydroxide, NaOH, is added to the buffer, the reaction is:

$$\underset{\text{Strong base}}{NaOH} + \underset{\text{Weak acid}}{H_2CO_3} \longrightarrow \underset{\text{Weak base}}{NaHCO_3} + \underset{\text{Water}}{H_2O}$$

The strong base has been changed to a __________ base. Should the pH change? __________

weak; no

it should not change it

152

Another buffer system in the body is the phosphate buffer system, which consists of NaH_2PO_4/Na_2HPO_4. What effect will the addition of a strong acid have on the pH of this buffer? ____________

no

153

Should the addition of base affect the pH of a phosphate buffer system?

____

no effect

154

What effect should the removal of acid (which is equivalent in effect to the addition of base) have on the pH of a buffer solution? ____________

none

155

The pH of the blood (a buffer solution) lies in the range of 7.35 to 7.45.

Normal metabolic processes add acid to the bloodstream. What effect should these acids have on the pH of the blood? ____________.

yes

156

Protein buffers are also present in the body. Should they behave similarly to the bicarbonate and phosphate buffers? ____________

Liquid Mixtures 5

SOLUTIONS

A solution consists of a *solute* (a relatively less abundant substance) dissolved in a *solvent* (a relatively more abundant substance). In living organisms the solvent is water.

157

a. When oxygen dissolves in water, oxygen is the ____________ (solute/solvent).

b. When alcohol is dissolved in water, alcohol is the ____________ (solute/solvent).

c. When salt is dissolved in water, water is the ____________ (solute/solvent).

a. solute
b. solute
c. solvent

158

A solution consists of a ____________ and a ____________.

solute; solvent

solution

159

Solute + solvent ⟶ ____________.

a. yes
b. homogeneous
c. transparent

160

Solutions are transparent and homogeneous. *Homogeneous* means "the same throughout," as compared with *heterogeneous*, which means "not the same throughout."

a. In a saltwater solution, is one part the same as any other part?

____

b. Solutions are ____________ (homogeneous/heterogeneous).

c. Solutions are ____________ (transparent/opaque).

a. yes
b. yes

161

In general, solutions pass through membranes.

a. If glucose is administered intravenously, will the glucose solution pass through the membranes in the body? ____________

b. Will saline (salt) solution pass through a membrane? ____________

water

162

For medical use, solutions are frequently described in terms of *percent* or *parts per hundred.*

In a percent solution, the percentage number indicates the number of grams of solute present in 100 mL of solution. The solvent is usually ____________.

163

a. A 5% glucose solution contains 5 g of glucose in ___________ mL of solution.

b. A 2% boric acid solution contains ___________ g of boric acid in ___________ mL of solution.

a. 100
b. 2, 100

164

Physiologic saline solution, 0.95% NaCl solution, contains ___________ g of NaCl in ___________.

0.95; 100 mL of solution

165

Another method of expressing solution concentration is in terms of *molarity*, or *moles* of solute per liter of solution. A *mole* of a compound is the number of grams of that compound equal to its molecular mass.

The atomic mass of hydrogen is 1 and that of oxygen, 16. The molecular mass of water, H_2O, which consists of 2 H's and 1 O is $(2 \times 1) + (1 \times 16)$, or ___________. Therefore, one mole of water has a mass of ___________ g.

18; 18

166

What is the molecular mass of glucose, $C_6H_{12}O_6$?___________
Atomic masses are: $C = 12, H = 1, O = 16$.

180 g

a. 1; 1 liter
b. 120; 1 liter
c. 0.1; solution
d. 5.85

167

A 1-molar (1M) glucose solution contains 1 mole (180 g) of glucose dissolved in enough water to make 1 liter of solution.

Use the following atomic masses in making calculations:

H	1	O	16
Mg	24	S	32
Na	23	Cl	35.5

a. To prepare a 1M $MgSO_4$ solution, take ____________ mole(s) of $MgSO_4$ and dissolve in enough water to prepare ____________ (how much?) of solution.

b. To prepare a 1M $MgSO_4$ solution, take ____________ g of $MgSO_4$ and dissolve in enough water to make ____________ (how much?) of solution.

c. To prepare a 0.1M NaCl solution, take ____________ mole(s) of NaCl and dissolve in enough water to make 1 liter of ____________.

d. To prepare a 0.1M NaCl solution, take ____________ g of NaCl and dissolve in enough water to make 1 liter of solution.

168

Osmolarity is a method of expressing solution concentration based on the number of particles in solution.

$$\text{osmolarity} = \text{molarity} \times \frac{\text{number of particles}}{\text{molecule}}$$

Osmolarity is expressed in the units osmoles, or *osmol.*

Consider the following reactions:

NaCl	$\longrightarrow$	$Na^+ + Cl^-$
1 molecule	$\longrightarrow$	2 particles
Glucose	$\longrightarrow$	glucose
1 molecule	$\longrightarrow$	1 particle
K_2SO_4	$\longrightarrow$	$2\,K^+ + SO_4^{2-}$
1 molecule	$\longrightarrow$	3 particles

a. A 1-molar (1M) glucose solution has a concentration of __________ osmol.

b. A 0.5M NaCl solution has a concentration of ____________ osmol.

c. A 0.4M K_2SO_4 solution has a concentration of ____________ osmol.

a. 1
b. 1
c. 1.2

169

Osmolarity may also be expressed in the units milliosmoles, or *mosmol*, where

$$1 \text{ osmol} = 1000 \text{ mosmol}$$

To change osmoles to milliosmoles, multiply by 1000.

a. A 0.15M NaCl solution has a concentration of ____________ osmol.

b. A 0.15M NaCl solution has a concentration of __________ mosmol.

Note that human body fluids have an osmolarity of approximately 300 milliosmol.

a. 0.30
b. 300

SUSPENSIONS

170

a. yes
b. no
c. no; the substance does not dissolve in water

When sand is placed in water and shaken, a sand-water suspension is produced.

a. When the shaking is stopped, does the sand settle? ____________

b. Does sand dissolve in water? ____________

c. In a suspension, is there a solute? ____________ Why or why not?
____

Note that particles in a suspension are so large that they settle under the influence of gravity.

171

a. heterogeneous
b. homogeneous
c. no
d. yes

a. A suspension is ____________ (homogeneous/heterogeneous).

b. A solution is ____________ (homogeneous/heterogeneous).

c. Is a suspension transparent? ____________

d. Is a solution transparent? ____________

172

yes; gravity

Blood is a suspension of blood cells in blood plasma. If blood is treated so that it doesn't clot and is then placed in a test tube, will the blood cells settle? ____________

What causes the blood cells to settle? ____________

COLLOIDS (COLLOIDAL DISPERSIONS)

Colloids consist of very tiny particles suspended in a liquid, usually water. (Colloids may also be suspended in solids and in gases.) Colloids differ from suspensions in that colloids do not settle. Colloidal dispersions are usually translucent or milky white. They do not pass through membranes.

173

One example of a colloid is protein.

a. A colloidal dispersion of protein will be ___________ (transparent/translucent).

b. A colloidal dispersion of protein ____________ (will/will not) settle.

c. A colloidal dispersion of protein ____________ (will/will not) pass through a membrane.

a. translucent
b. will not
c. will not

174

Many proteins are colloids. Should such proteins be able to pass through the membranes of the kidneys and be present in the urine? ___________

Should salt, NaCl? ____________

no; yes

PART II
Organic Chemistry

The Covalent Bond 6

175

You've already learned that atoms tend to lose electrons to form positively charged ions if they have 1, ____________, or ____________ electrons in their valence shells.

On the other hand, atoms that have ____________ or ____________ electrons in their valence shells tend to gain electrons to form _________ charged ions.

2; 3; 6; 7; negatively

176

Atoms with 4 or 5 electrons in their valence shells tend to *share* electrons. When atoms share electrons, they form a *covalent bond.*

Which of the following illustrated atoms would form covalent bonds?

____

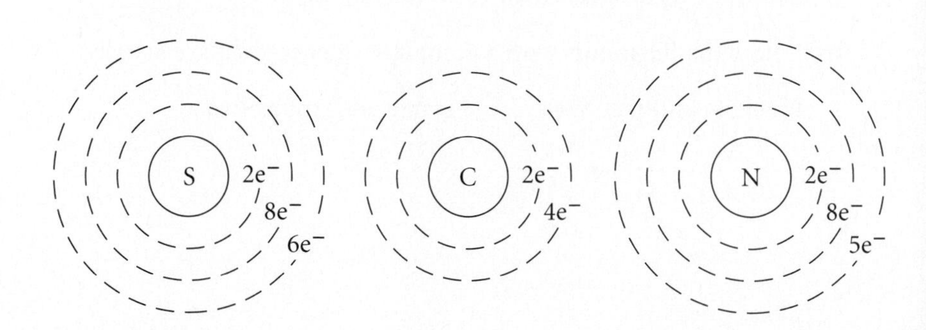

C and N

a. Mg, H, Al
b. Cl, S
c. Ne
d. stable and unreactive

177

Answer the following questions about these illustrated atoms.

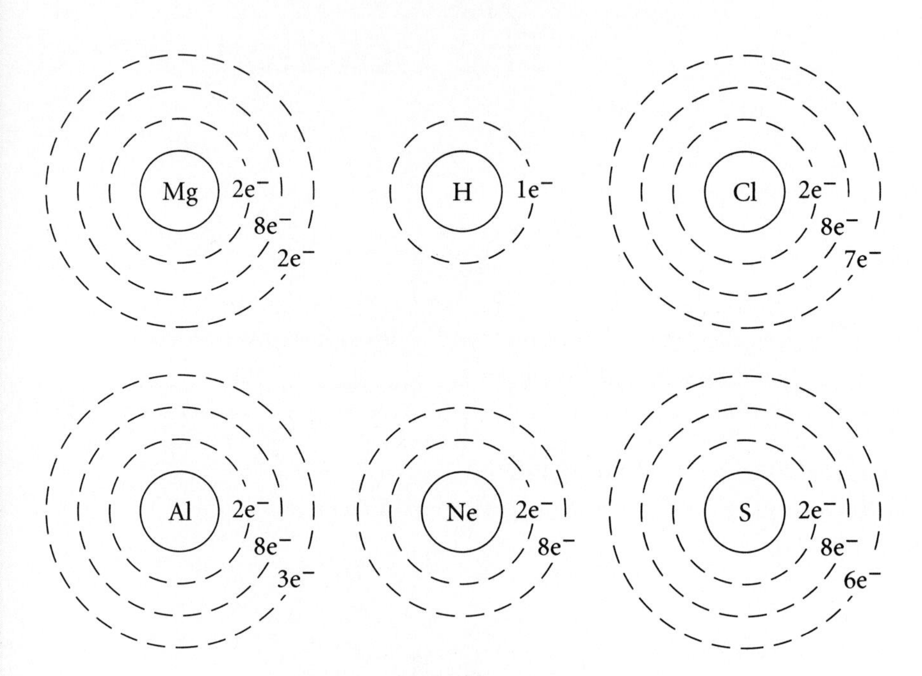

a. Which of the atoms will lose electron(s) to form positively charged ions? ____________

b. Which of the atoms will gain electron(s) to form negatively charged ions? ____________

c. Which of the atoms has a complete valence shell? ____________

d. You know that elements with a complete valence shell are usually:

__ unstable and reactive __ stable and unreactive

178

a. When two atoms share electrons, they are held together by a ____________ bond.

b. When two atoms share electrons, no ions are produced. Is the resulting compound an electrolyte or a nonelectrolyte? ___________

a. covalent
b. nonelectrolyte

179

Atoms with 4 or 5 electrons in their valence shells tend to share electrons, but under some conditions atoms with more or fewer electrons in their valence shells may also share electrons.

When one atom loses an electron and another atom gains that electron, the ions thus formed are held together by a(n) ___________ (covalent/ionic) bond.

ionic

180

a. Compounds containing ionic bonds are ____________ (electrolytes/nonelectrolytes).

b. Compounds containing only covalent bonds are ____________ (electrolytes/nonelectrolytes).

a. electrolytes
b. nonelectrolytes

4

181

The carbon atom, atomic number 6, has 4 electrons in its valence shell. It thus needs ____________ more electrons to become stable. It can get these electrons by sharing. The 4 electrons in the valence shell of the carbon atom may be represented as here, with one dot for each electron in the valence shell.

$$\cdot\overset{\cdot}{\underset{\cdot}{C}}\cdot$$

2

182

When atoms share electrons, they always try to reach stable configurations of 8 electrons in the valence shells. The only exception to this rule is the hydrogen atom, which reaches a stable configuration when it has 2 electrons in its valence shell, which is also its only energy level. This first energy level can hold only ____________ electrons.

4

183

Consider the carbon atom with 4 electrons in its valence shell. With how many hydrogen atoms can it share electrons to reach a stable configuration of 8 electrons in its valence shell? ____________

8; 2

184

The four hydrogen atoms can share electrons with a carbon atom to form a compound of the following type:

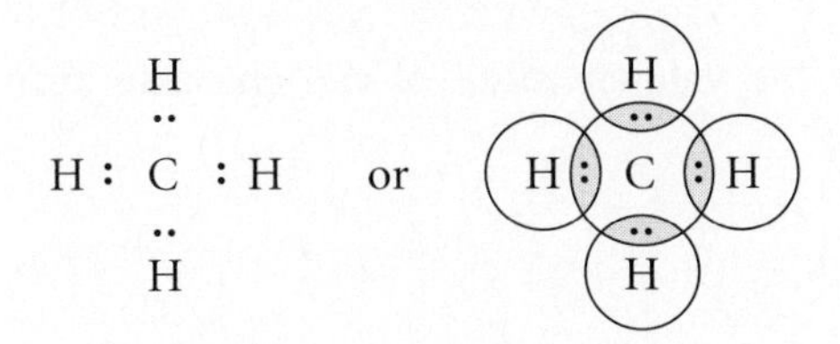

There are ____________ electrons around the carbon atom and ____________ electrons around each hydrogen atom.

2

185

When two hydrogen atoms combine, they share electrons. Each hydrogen atom then has ____________ electrons around it.

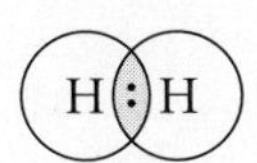

4

186

In the following compound, what is the total number of covalent bonds? ____________

H
‥
H : C : H
‥
H

7

187

The chlorine atom, atomic number 17, has ____________ electrons in its valence shell.

The electrons in the valence shell of the chlorine atom may be represented as

$$\cdot\ \overset{\cdot\cdot}{\underset{\cdot\cdot}{\mathrm{Cl}}}\ :$$

When two chlorine atoms combine, they share electrons to reach a stable configuration of 8 electrons in the valence shell of each atom, or

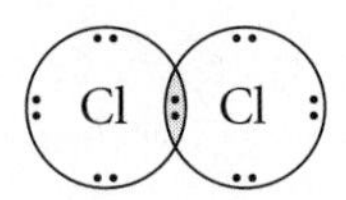

covalent

188

What type of bond is there between two chlorine atoms? ____________

189

A covalent bond is frequently indicated by a short line rather than by dots; thus, the hydrogen molecule may be represented as H : H or H—H. The short line indicates a pair of shared electrons.

The compound CH_4 may be represented as

$$\mathrm{H} : \overset{\overset{\mathrm{H}}{\cdot\cdot}}{\underset{\underset{\mathrm{H}}{\cdot\cdot}}{\mathrm{C}}} : \mathrm{H} \quad \text{or} \quad \mathrm{H}-\overset{\overset{\mathrm{H}}{|}}{\underset{\underset{\mathrm{H}}{|}}{\mathrm{C}}}-\mathrm{H}$$

Complete the second representation of the chlorine molecule, Cl_2.

```
   ..   ..
 : Cl : Cl :  or
   ..   ..
```

Cl—Cl

190

a. The carbon atom, atomic number 6, needs how many electrons to complete its valence shell? ____________

b. Because the carbon atom has 4 electrons in its valence shell, and because it will tend to share these electrons, how many covalent bonds will a carbon atom form? ____________

a. 4
b. 4

The carbon atom, therefore, must have four covalent bonds, or must have four bonds attached to it. These bonds may be indicated as follows:

```
  |       |
 —C—     —C═     —C≡     ═C═
  |
```

Note that each carbon atom has four bonds attached to it, regardless of how the bonds are arranged.

191

The hydrogen atom has 1 electron in its outer energy level, the first energy level. When the hydrogen atom shares electrons, how many more does it need to complete that valence shell? ____________

1

a. 6
b. 2
c. 2

192

a. The oxygen atom, atomic number 8, has ____________ electrons in its valence shell.

b. When the oxygen atom shares electrons, how many more electrons does it need to complete its valence shell? ____________

c. Therefore, each oxygen atom must have how many bonds attached to it? ____________

```
    H
    |
H — C — H
    |
    H
```

193

Draw the structure of the compound formed when four hydrogen atoms form bonds with a central carbon atom.

C

a. 3
b. 3

194

The arrangement of carbon atoms C—C indicates a bond or a shared electron pair between two carbon atoms.

a. How many hydrogen atoms may be attached to the left carbon atom? ____________

b. How many hydrogen atoms may be attached to the right carbon atom? ____________

195

a. Diagram all the hydrogen atoms attached to this structure:

C—C

b. How many bonds does each carbon atom have? ____________

c. How many bonds does each hydrogen atom have? ____________

a.

```
    H   H
    |   |
H—C—C—H
    |   |
    H   H
```

b. 4
c. 1

196

Consider this arrangement of carbon atoms: C—C—C

a. How many hydrogen atoms may be attached to the left carbon atom? ____________

b. How many hydrogen atoms may be attached to the center carbon atom? ____________

c. How many hydrogen atoms may be attached to the right carbon atom? ____________

a. 3
b. 2
c. 3

197

Diagram the compound containing three attached carbon atoms, indicating all the hydrogen atoms connected to those carbon atoms.

C—C—C

```
    H   H   H
    |   |   |
H—C—C—C—H
    |   |   |
    H   H   H
```

a. 3
b. 3
c. 3
d. 3
e. 0 (or none)

198

Consider this arrangement of carbon atoms:

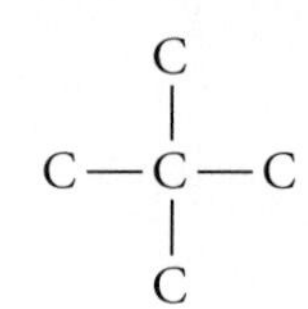

a. How many hydrogen atoms may be attached to the right carbon atom? ____________

b. How many hydrogen atoms may be attached to the left carbon atom? ____________

c. How many hydrogen atoms may be attached to the upper carbon atom? ____________

d. How many hydrogen atoms may be attached to the lower carbon atom? ____________

e. How many hydrogen atoms may be attached to the center carbon atom? ____________

a.

```
          H
          |
       H—C—H
   H      |      H
   |      |      |
H—C——C——C—H
   |      |      |
   H      |      H
       H—C—H
          |
          H
```

b. It already has four bonds.

199

a. Diagram the structure of the compound containing the following arrangement of carbon atoms, indicating all the hydrogen atoms.

```
    C
    |
C—C—C
    |
    C
```

b. Why can there be no hydrogen atoms attached to the center carbon atom?____

200

Consider the following arrangement of carbon atoms: C═C.

There are two bonds (a double bond) between the carbon atoms. Each bond (a single bond) represents 1 pair of shared electrons. A double bond represents ______________________________.

2 pairs of shared electrons

201

a. To become stable, a carbon atom must form how many covalent bonds? ____________

b. In the structure C═C, how many hydrogen atoms may be attached to the left carbon atom? ____________

c. In the space provided, diagram the structure of the compound containing two carbon atoms connected by a double bond. Show all of the hydrogen atoms.

When you have completed your diagram, check it by counting the number of bonds around each carbon atom.

a. 4
b. 2
c.

```
    H   H
    |   |
H — C = C — H
```

a. 2
b. 1
c. 3

202

Consider the following arrangement of carbon atoms: C=C—C

a. How many hydrogen atoms may be attached to the left carbon atom? ____________

b. How many hydrogen atoms may be attached to the center carbon atom? ____________

c. How many hydrogen atoms may be attached to the right carbon atom? ____________

```
   H   H   H
   |   |   |
H—C=C—C—H
           |
           H
```

203

Complete the structure of the following compound, showing all the hydrogen atoms.

C=C—C

```
     Cl
     |
Cl—C—Cl
     |
     Cl
```

204

Diagram the structure of carbon tetrachloride, CCl_4, in which each chlorine atom has one bond attached to it.

205

Diagram the structure of chloroform, $CHCl_3$.

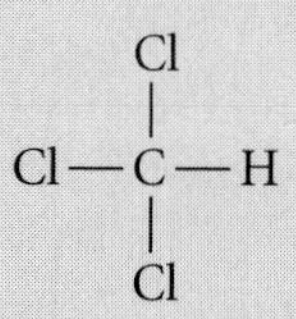

```
      Cl
      |
Cl — C — H
      |
      Cl
```

206

Nitrogen, atomic number 7, has 5 electrons in its valence shell and so must have three bonds attached to it. Diagram the ammonia molecule, NH_3.

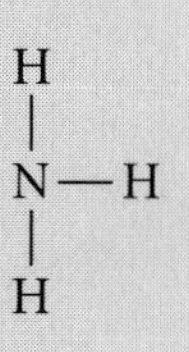

```
H
|
N — H
|
H
```

Note that the following structure

```
     H   H   H   H
     |   |   |   |
 H — C — C — C — C — H
     |   |   |   |
     H   H   H   H
```

may be simplified and written as $CH_3CH_2CH_2CH_3$. Likewise,

```
     H   H   H   H   H
     |   |   |   |   |
 H — C — C — C — C — C — H
     |   |   |   |   |
     H   |   H   H   H
         |
     H — C — H
         |
         H
```

may be simplified and written as $CH_3CH(CH_3)CH_2CH_2CH_3$, and

```
     H   H   H   H
     |   |   |   |
 H — C — C = C — C — H
     |           |
     H           H
```

may be simplified and written as $CH_3CH{=}CHCH_3$.

207

How may the following structures be written in simplified form?

a.

```
    H   H
    |   |
H — C — C — H
    |   |
    H   H
```

____

b.

```
    H   H   H   H   H
    |   |   |   |   |
H — C — C — C — C — C — H
    |   |   |   |   |
    H   H   |   H   H
            |
        H — C — H
            |
            H
```

____

c.

```
    H
    |
H — C — C ≡ C — H
    |
    H
```

____

d.

```
             H
             |
         H — C — H
             |
    H        |        H   H
    |        |        |   |
H — C ——————— C ——————— C — C — H
    |        |        |   |
    H        |        H   H
             |
         H — C — H
             |
             H
```

____

a. CH_3CH_3
b. $CH_3CH_2CH(CH_3)CH_2CH_3$
c. $CH_3C \equiv CH$
d. $CH_3C(CH_3)_2CH_2CH_3$

The structure for benzene, C_6H_6, may be represented in two different ways, depending on the arrangement of the electrons:

H H
H—C C—H ⟷ H—C C—H
H—C C—H H—C C—H
H H

Such structures are called *resonance* structures and are indicated by a double-headed arrow.

Often, either of these resonance structures is used to represent benzene. However, for simplicity, benzene is usually abbreviated as

a. —CH_3

208

What is the abbreviated structure for the following?

a.

H H
C—H
H—C C
H
H—C C—H
H

____

b.

```
          H
          |
          C
        //  \      OH
    H—C       C  /
      |       ||  H   H
      |       ||  |   |
    H—C       C—C—C—H
        \\  /     |   |
          C       H   H
          |
          H
```

____

209

Draw the complete structure for each of the following simplified formulas:

a. $CH_3CH_2CH_3$

b. $CH_3CH{=}CHCH_3$

c. (benzene ring)—Cl

d. $CH_3CH(CH_3)CH_2CH_2CH_3$

---

b. (benzene ring) with OH and CH_2CH_3

a.

```
   H  H  H
   |  |  |
H—C—C—C—H
   |  |  |
   H  H  H
```

b.

```
   H  H  H  H
   |  |  |  |
H—C—C=C—C—H
   |        |
   H        H
```

c.

```
       H
       |
       C
     //  \
  H—C     C—Cl
    |     ||
  H—C     C—H
     \\  /
       C
       |
       H
```

or

```
       H
       |
       C
     /  \\
  H—C     C—Cl
    ||    |
  H—C     C—H
     \  //
       C
       |
       H
```

d.

```
   H  H  H  H  H
   |  |  |  |  |
H—C—C—C—C—C—H
   |  |  |  |  |
   H  |  H  H  H
      |
   H—C—H
      |
      H
```

Polar and Nonpolar Covalent Bonds 7

210

Electronegativity is the attraction of an atom for electrons. The following table indicates the electronegativities of several elements.

F 4.0	Br 2.8	Al 1.5
O 3.5	C 2.5	Ca 1.0
Cl 3.1	H 2.1	Na 0.9
N 3.0		

Note that nonmetals have higher electronegativities than do metals. The difference in atoms' electronegativities determines the type of bond that can form between them.

When atoms of two nonmetals combine, the difference in electronegativities is small or zero and a covalent bond results.

When atoms of a metal and a nonmetal react, the difference in electronegativities is relatively larger and an ionic bond results.

a. Which of these elements has the greatest attraction for electrons?

____

b. Which of these elements has the least attraction for electrons?

____

a. F
b. Na

211

In the compound Cl—Cl or Cl:Cl, there is a pair of shared electrons between the chlorine atoms. Is there a difference in electronegativity between the two chlorine atoms?____________

no

a. Cl
b. closer to

212

Because the two chlorine atoms have the same electronegativity, the electrons should be shared equally between them. That is, the electrons should not be closer to one chlorine atom than to the other. Such a bond is called a *nonpolar covalent bond*, and such a compound is called a *nonpolar compound*. Nonpolar means there is no negative end and no positive end to the bond or to the molecule.

a. In the covalent compound HCl, H—Cl, or H:Cl, which element has the greater electronegativity? ____________

b. Because Cl is more electronegative than H, the shared electrons should be ____________ (closer to/farther from) the chlorine.

a. yes
b. polar
c. polar

213

Because the shared electrons are closer to the Cl than to the H, the Cl will have a partial negative charge and the H will have a partial positive charge. This may be indicated by a δ (Greek delta) sign.

$$\overset{\delta+}{H}\overset{\delta-}{Cl}$$

Note that this is a partial charge, not an ionic charge.

a. Is one end of the molecule more negative than the other end? ____________

b. The covalent bond between the H and the Cl is ____________ (polar/nonpolar).

c. Is the HCl molecule polar or nonpolar? ____________

nonpolar

214

Consider the covalent compound XY. If both atoms have the same electronegativity, will the covalent bond between them be polar or nonpolar? ____________

215

If there is a difference in electronegativity between X and Y, the covalent bond between them will be ____________ and the compound XY will be ____________.

polar; polar

216

Now consider the covalent compound CO_2, or O::C::O.

a. Which atom is more electronegative? ____________ Which is less electronegative? ____________

b. Which atom will have a partial + charge? ____________ Which will have a partial − charge? ____________

a. O; C
b. C; O

217

The CO_2 molecule may be represented as a linear molecule.

$$\overset{\delta-}{O}::\overset{\delta+}{C}::\overset{\delta-}{O}$$

Both ends of the CO_2 molecule have a partial ____________ (positive/negative) charge.

negative

218

a. Does the CO_2 molecule have a positive and a negative end?

____

b. Is CO_2 a polar or nonpolar compound? ____________

a. no
b. nonpolar

nonpolar

219

In general, if a covalent compound is symmetrical, it will be __________ (polar/nonpolar).

H; O

220

In the H_2O molecule, which contains covalent bonds, which atom has a partial positive charge? ____________

Which has a partial negative charge? ____________

b; a

221

The structure of the water molecule may be represented as either

$$\text{a. } \overset{\delta+}{H}—\overset{\delta-}{O}—\overset{\delta+}{H} \quad \text{or} \quad \text{b. } \overset{\delta+}{H}—\overset{\delta-}{O} \diagdown H^{\delta+}$$

The structure in a is theoretical; it does not exist.

Which structure, a or b, represents a polar compound? ___________

Which represents a nonpolar compound? ____________

b

222

The water molecule has been found to be a polar molecule. Which of the structures in the preceding question more accurately depicts the water molecule? ____________

223

One general rule of solubility is that *like dissolves like*. That is, polar liquids dissolve polar compounds, and nonpolar liquids dissolve nonpolar compounds. Is water a polar or a nonpolar liquid? __________

polar

224

a. Should NaCl, a polar compound, dissolve in water? ____________

b. Should benzene, a nonpolar compound, dissolve in water? ____________

c. Should acetone, a nonpolar liquid, dissolve in benzene, a nonpolar liquid? ____________

a. yes
b. no
c. yes

Hydrogen Bonds 8

225

Consider two water molecules whose structures and partial charges are:

$$\overset{\delta+}{H}-\overset{\delta-}{O}\!\diagdown H^{\delta+} \quad \text{and} \quad \overset{\delta+}{H}-\overset{\delta-}{O}\!\diagdown H^{\delta+}$$

Which partial charges should attract one another? ____________________

____

____

____

____

the negative charge of one water molecule and the positive charge of the other; the positive charge of one water molecule and the negative charge of the other.

weak

226

Water molecules can form a weak bond between the H of one molecule and the O of another. Such a bond, called a *hydrogen bond*, is caused by the attraction of the partially positive H of one water molecule and the partially negatively charged O (more electronegative) of another water molecule. Hydrogen bonds may be indicated by dotted lines:

```
              H
             /
        H—O···H
          :     \
         H       O···H
          \     /     \
           O···H       O—H    H
          /            :   ·. /
         H             H     O
          :             \  .· \
        H—O···H          O—H   H
         :  \   \         :
      H—O···H   O—H      H
          \      :       /
           H    H       O
                 \    .· \
                  O···H   H
                 /     \ .·
                H       O
                       /
                      H
```

Is a hydrogen bond a strong or a weak bond? ____________

hydrogen bonds

227

If many water molecules are bonded together in a similar manner, what type of bond should hold them together? ____________

228

Hydrogen bonds can also occur between the H attached to an N and an O that is part of a C=O group. Such hydrogen bonds are present in proteins and in DNA.

Note the following diagram, which illustrates a very small part of a protein molecule.

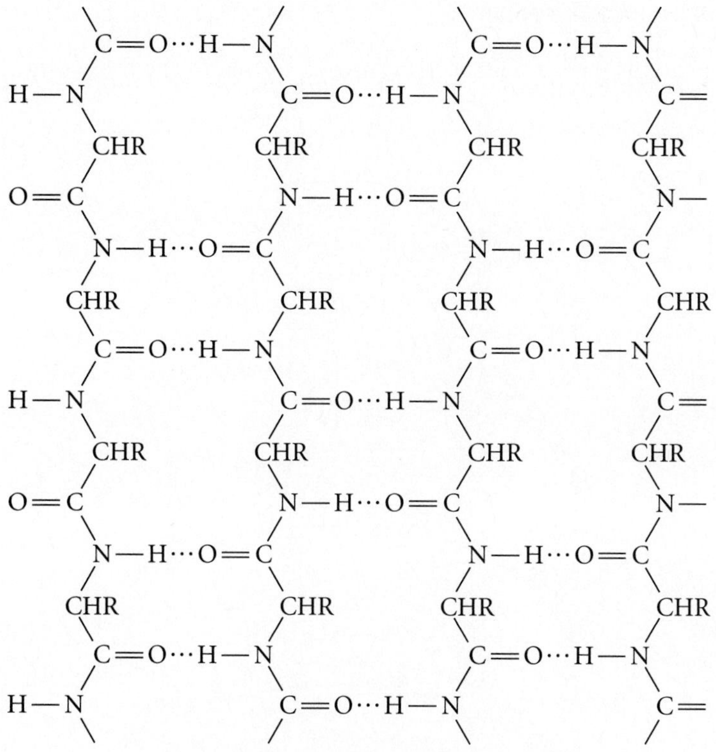

```
       \           /                 \           /
        C=O···H—N                     C=O···H—N
       /           \                 /           \
  H—N               C=O···H—N                      C=
       \           /           \                 /
        CHR      CHR             CHR           CHR
       /           \            /                 \
  O=C               N—H···O=C                      N—
       \           /           \                 /
        N—H···O=C                N—H···O=C
       /           \            /                 \
      CHR           CHR      CHR                    CHR
       \           /           \                 /
        C=O···H—N                 C=O···H—N
       /           \             /              \
  H—N               C=O···H—N                      C=
       \           /           \                 /
        CHR      CHR             CHR           CHR
       /           \            /                 \
  O=C               N—H···O=C                      N—
       \           /           \                 /
        N—H···O=C                N—H···O=C
       /           \            /                 \
      CHR           CHR      CHR                    CHR
       \           /           \                 /
        C=O···H—N                 C=O···H—N
       /           \             /              \
  H—N               C=O···H—N                      C=
       \           /           \                 /
```

Pleated sheet structure of a protein

What type of bond holds the sections together? ____________

hydrogen bonds

229

The hydrogen bonds in the diagram in problem 228 exist between the H attached to a(n) ____________ atom and the =O attached to a(n) ____________ atom.

N; C

hydrogen bonds

230

We have seen that hydrogen bonding accounts for the pleated sheet structure of some proteins.

Some proteins have a helical structure, as shown in the following diagram:

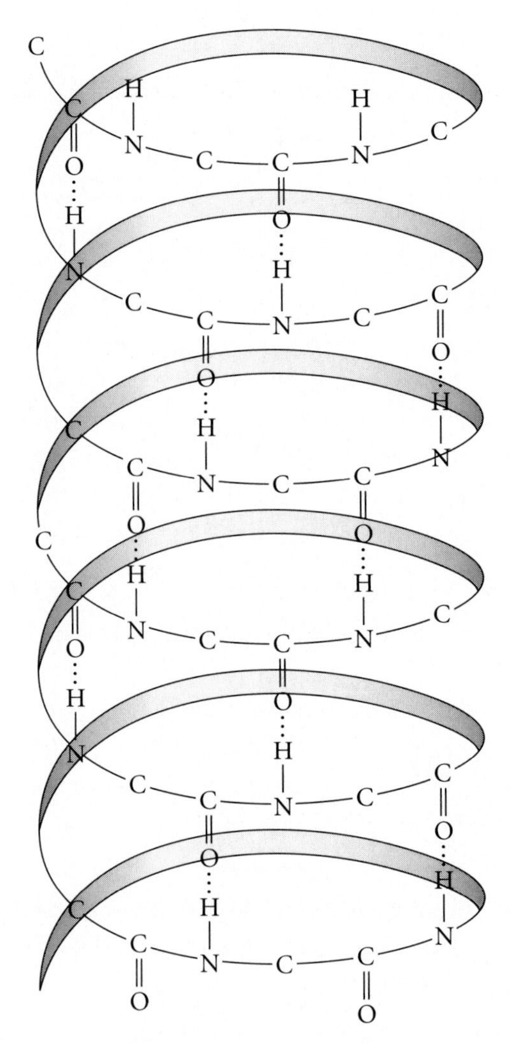

Helical structure of a protein

What type of bond holds the coil in its designated shape? ____________

231

In both the pleated sheet and helical structures of proteins, the hydrogen bonds present exist between the H attached to a(n) ____________ and the O that is part of a(n) ____________ group.

N; C=O

232

Consider the double helical shape of DNA in the following diagram.

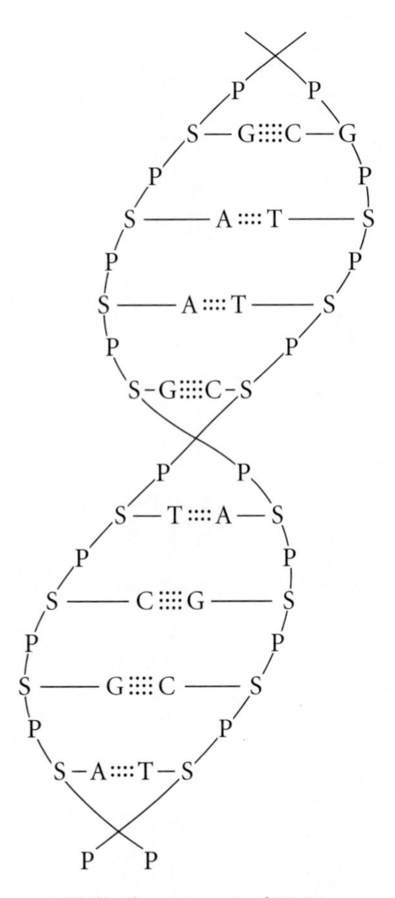

Helical structure of DNA

What type of bond holds the two halves together? ____________

hydrogen bonds

2; hydrogen bonds

233

The DNA structure contains the following substances: adenine (A), guanine (G), cytosine (C), and thymine (T).

How many bonds are present between A and T? ____________

What type of bonds are these? ____________

3; hydrogen bonds

234

How many bonds are present between C and G? ____________

What type of bonds are these? ____________

weak

235

When the DNA molecule replicates, the two halves come apart, breaking the bonds holding them together.

Are hydrogen bonds strong or weak bonds? ____________

easy

236

Should it be easy or difficult to "open up" the DNA chain? ____________

Functional Groups in Organic Compounds 9

Functional groups impart certain sets of properties to compounds. One example of such a functional group is the —*OH group*.

ALCOHOLS

237

The —OH group in an ionic compound is called a hydroxide ion.

$$\text{NaOH} \longrightarrow \text{Na}^+ + \text{OH}^-$$

Ionic

The —OH group in a covalent compound is called an *alcohol* group; it does not ionize.

```
      H
      |
  H—C—OH
      |
      H
```

Covalent, not ionized

Is a solution of NaOH an electrolyte? ____________

yes

no

238

Consider this compound:

```
      H
      |
  H — C — OH
      |
      H
```

Is a solution of this compound an electrolyte? ____________

The —OH group has the bonds —O—H. It is frequently written as —OH; the bond between the hydrogen and the oxygen is understood and is not written. Note that the oxygen atom still has two bonds attached to it.

In general, covalent compounds containing one or more —OH groups are called *alcohols*.

all of them

239

Which of the following compounds is(are) an alcohol? ____________

a.
```
      H   H   H
      |   |   |
  H — C — C — C — OH
      |   |   |
      H   H   H
```

b.
```
              H
              |
          H — C — H
              |
          H   |   H
          |   |   |
      H — C — C — C — H
          |   |   |
          H   OH  H
```

c. $CH_3CH(OH)CH_3$

d. $CH_3CH_2CH_2CH(CH_3)CH_2OH$

e. (benzene ring bearing CH_2CH_2OH and CH_3 on adjacent positions)

$-CH_2CH_2OH$

$-CH_3$

240

Glycerol, $C_3H_8O_3$, is one of the primary constituents of fats. Its structure is:

```
     H   H   H
     |   |   |
 H — C — C — C — H
     |   |   |
     OH  OH  OH
```

Is glycerol an alcohol? ____________ Why or why not? ____________

yes; it contains —OH groups

241

The names of alcohols are written either with the word *alcohol* at the end or with the ending *ol.* CH_3OH is methyl alcohol, or methanol. CH_3CH_2OH is ethyl alcohol, or ethanol.

If the name of a compound ends in *ol*, that compound is a(n) ________.

alcohol

242

Which of the following compounds is(are) an alcohol? ____________

a. cholesterol

b. tocopherol (vitamin E)

c. thiamine (vitamin B_1)

d. cortisone

e. retinol (vitamin A)

a, b, e

ALDEHYDES

A second functional group is the *aldehyde* group. Aldehydes contain a —CHO group. Note that in the *aldehyde* group the oxygen atom has a double bond attached to it.

```
   H
   |
 —C=O
```

aldehyde

243

The compound following is an:

```
     H   H   H
     |   |   |
 H—C—C—C=O
     |   |
     H   H
```

__ aldehyde __ alcohol

The previous compound may also be written as CH_3CH_2CHO, where the —CHO group indicates an aldehyde. Note that the aldehyde group is at the end of the molecule.

Check the correct name for each of the compounds in the next five problems, and explain why you chose the answer you did.

244

```
     H   H   H   H
     |   |   |   |
 H — C — C — C — C — OH
     |   |   |   |
     H   H   H   H
```

__ alcohol __ aldehyde __ some other type

Why? ____________________

alcohol; it has an —OH group

245

```
     H   H
     |   |
 H — C — C = O
     |
     H
```

__ alcohol __ aldehyde __ some other type

Why? ____________________

aldehyde; it has a —CHO group (or equivalent answer)

some other type; it has neither an —OH group nor a —CHO group

246

```
    H       H
    |       |
H — C — C — C — H
    |   ‖   |
    H   O   H
```

__ alcohol __ aldehyde __ some other type

Why? ______________________________

aldehyde; it has a —CHO group

247

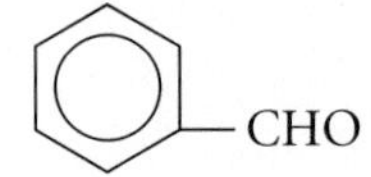

__ alcohol __ aldehyde __ some other type

Why? ______________________________

alcohol; it has an —OH group

248

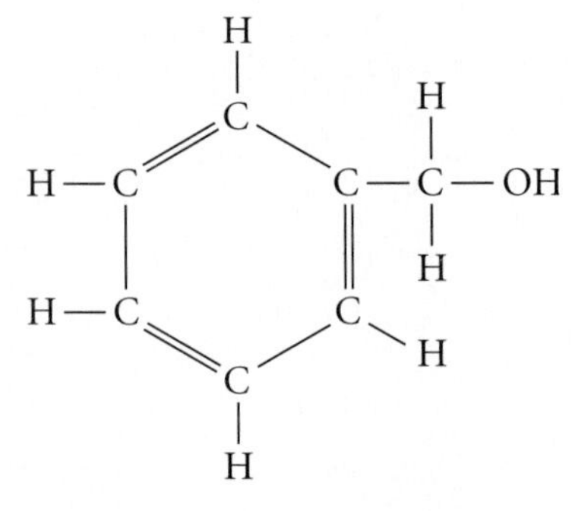

__ alcohol __ aldehyde __ some other type

Why? ______________________________

249

Aldehydes are named either with the word *aldehyde* at the end or with the ending *al*.

$$\begin{array}{c} H \\ | \\ HC{=}O \end{array}$$

is formaldehyde, or methanal.

CH_3CHO is acetaldehyde, or ethanal.

Retinal is a compound involved in the visual cycle. What kind of compound is it? ____________ Why? ____________________________

an aldehyde; its name ends in *al*

250

a. The formula for an alcohol group is ____________.

b. The formula for an aldehyde group is ____________.

a. —OH
b. —CHO

KETONES

Ketones are another functional group. A ketone contains a carbon atom that is attached to a double-bonded oxygen atom and is between two other carbon atoms.

$$\begin{array}{c} —C—C—C— \\ \| \\ O \end{array}$$

An example of a ketone is acetone, found in the bloodstream and the urine of diabetics.

$$\begin{array}{c} CH_3—C—CH_3 \\ \| \\ O \end{array}$$

Identify the compounds in the next nine problems as alcohols, aldehydes, or ketones.

alcohol

251

```
    H  H  H  H  H
    |  |  |  |  |
 H—C—C—C—C—C—H
    |  |  |  |  |
    H  H  H  OH H
```

__ alcohol __ aldehyde __ ketone

aldehyde

252

```
    H  H  H  H
    |  |  |  |
 H—C—C—C—C=O
    |  |  |
    H  H  H
```

__ alcohol __ aldehyde __ ketone

ketone

253

```
    H  H     H  H
    |  |     |  |
 H—C—C—C—C—C—H
    |  |  ‖  |  |
    H  H  O  H  H
```

__ alcohol __ aldehyde __ ketone

254

$CH_3CH(OH)CH_3$

__ alcohol __ aldehyde __ ketone

alcohol

255

```
        H     H
        |     |
        C-----C
   H  / |     | \
   | /  H     H  \
   C               C=O
   | \  H     H  /
   H  \ |     | /
        C-----C
        |     |
        H     H
```

__ alcohol __ aldehyde __ ketone

ketone

256

C_6H_5—C—CH_3 (C=O)

```
      C—CH3
      ||
      O
```

__ alcohol __ aldehyde __ ketone

ketone

aldehyde and ketone

257

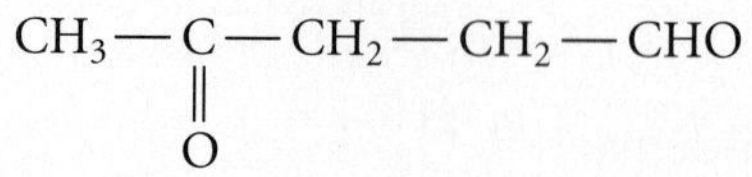

$$CH_3-\underset{\underset{O}{\|}}{C}-CH_2-CH_2-CHO$$

__ alcohol __ aldehyde __ ketone

alcohol and aldehyde

258

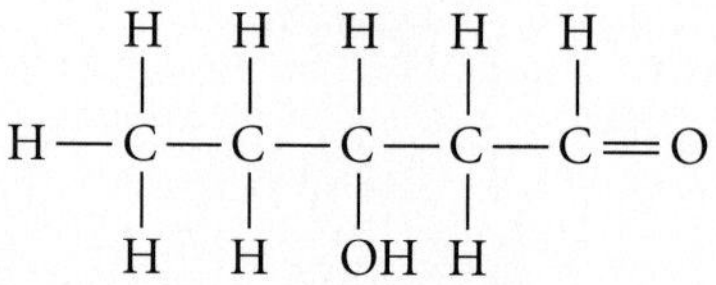

$$H-\underset{H}{\overset{H}{C}}-\underset{H}{\overset{H}{C}}-\underset{OH}{\overset{H}{C}}-\underset{H}{\overset{H}{C}}-\overset{H}{C}=O$$

__ alcohol __ aldehyde __ ketone

aldehyde and ketone

259

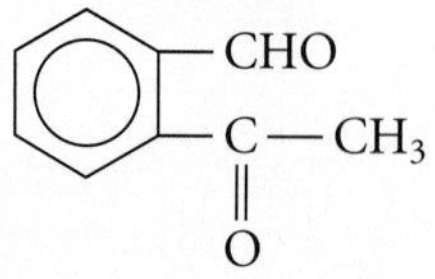

$$C_6H_4(CHO)(\underset{\underset{O}{\|}}{C}-CH_3)$$

__ alcohol __ aldehyde __ ketone

260

The names of most ketones end in *one*. An example of a ketone is acetone. Which of the following substances are ketones? ____________
Alcohols? ____________ Aldehydes? ____________

a. corticosterone b. pyridoxal (a B-vitamin)
c. cortisol d. androsterone
e. estriol f. pentanal
g. pregnandiol

ketones: a, d
alcohols: c, e, g
aldehydes: b, f

ORGANIC ACIDS

261

A fourth functional group is the *carboxyl* group, —COOH or

$$-\overset{\overset{\displaystyle O}{\|}}{C}-OH$$

The carboxyl group ionizes as follows:

$$-\overset{\overset{\displaystyle O}{\|}}{C}-OH \longrightarrow -\overset{\overset{\displaystyle O}{\|}}{C}-O^- + H^+$$

Because the carboxyl group yields H^+ ions, it is a(n) ____________.

Because organic acids contain a carboxyl group, they are also called *carboxylic acids.*

acid

a. an aldehyde
b. an acid

262

The —COOH group may also be written as

$$-\overset{\overset{\displaystyle O}{\|}}{C}-OH \quad \text{or} \quad -\overset{\overset{\displaystyle OH}{|}}{C}=O$$

a. A compound with a —CHO group is called ____________.

b. A compound with a —COOH group is called ____________.

carboxyl, or —COOH

263

Citric acid, α-ketoglutaric acid, succinic acid, and oxaloacetic acid are a part of the Krebs cycle, the human body's primary metabolic cycle. What functional group should they all have in common? ____________

a. ketone and carboxyl (organic acid)
b. alcohol and carboxyl

264

What functional groups are present in the following compounds?

a. pyruvic acid, produced during the oxidation of glucose ____________

$$CH_3-\underset{\underset{\displaystyle O}{\|}}{C}-COOH$$

b. lactic acid, formed during the fermentation of milk ____________

$$CH_3-\underset{\underset{\displaystyle OH}{|}}{CH}-COOH$$

c. tartaric acid, found in grapes ____________

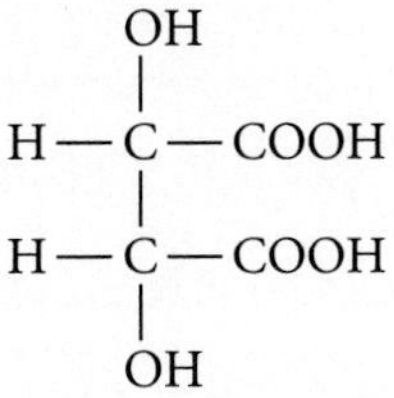

$$
\begin{array}{c}
\text{OH} \\
| \\
\text{H}-\text{C}-\text{COOH} \\
| \\
\text{H}-\text{C}-\text{COOH} \\
| \\
\text{OH}
\end{array}
$$

c. **alcohol and carboxyl**

AMINES

A compound that contains an NH_2 functional group is called an *amine*. It is written as $-NH_2$.

265

Which of the following compounds are amines? ____________

a. CH_3NH_2

b. 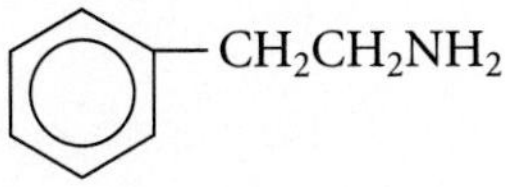 (benzene ring)$-CH_2CH_2NH_2$

c. 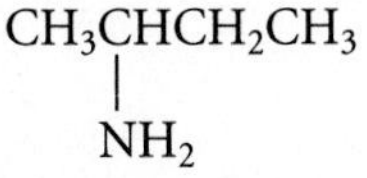 $\underset{\displaystyle NH_2}{CH_3\underset{|}{C}HCH_2CH_3}$

d. 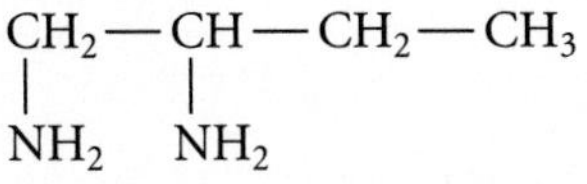 $\underset{NH_2}{\underset{|}{CH_2}}-\underset{NH_2}{\underset{|}{CH}}-CH_2-CH_3$

all of them

a. —COOH (carboxyl)
b. an amine
c. an alcohol
d. an aldehyde
e. a ketone

266

a. An organic acid is a compound containing a ____________ group.

b. A compound containing an $—NH_2$ group is called ____________.

c. A compound containing an —OH group is called ____________.

d. A compound containing a —CHO group is called ____________.

e. A compound containing a $C—\underset{\underset{O}{\|}}{C}—C$ group is called ____________.

Identify the types of compounds in the next five problems. Write out the reasons for your choices.

amine: it contains an $—NH_2$ group

267

```
    H   H   H
    |   |   |
H — C — C — C — H
    |   |   |
    H  NH2  H
```

__ alcohol __ aldehyde
__ amine __ acid
__ ketone

Why? ____

268

```
    H   H   H   H
    |   |   |   |
H — C — C — C — C = O
    |   |   |
    H   OH  H
```

__ alcohol and acid __ alcohol and aldehyde
__ amine and alcohol __ ketone and alcohol

Why? ____________________

alcohol and aldehyde; it contains both —CHO and —OH groups

269

```
    H   H   H       H
    |   |   |       |
H — C — C — C — C — C — H
    |   |   |   ‖   |
    H   OH  H   O   H
```

__ amine and alcohol __ alcohol and aldehyde
__ acid and alcohol __ ketone and alcohol

Why? ____________________

ketone and alcohol; it contains both —OH and C—C—C groups (with =O on the central C)

```
C — C — C
    ‖
    O
```

270

```
    H   H   H    H   H
    |   |   |    |   |
H — C — C — C  — C — C — OH
    |   |   |    |   |
    H   H   NH₂  H   H
```

Type ____________________
Why? ____________________

amine and alcohol, or amino alcohol; it contains both $—NH_2$ and —OH groups

amine and organic acid; it contains both $—NH_2$ and $—COOH$ groups

271

$$
\begin{array}{ccccccccc}
 & & H & & H & & H & & O \\
 & & | & & | & & | & & \| \\
H & — & C & — & C & — & C & — & C — OH \\
 & & | & & | & & | & & \\
 & & H & & NH_2 & & H & &
\end{array}
$$

Type ____________________

Why? ____________________

Identify the types of compounds in the next four problems.

alcohol and organic acid

272

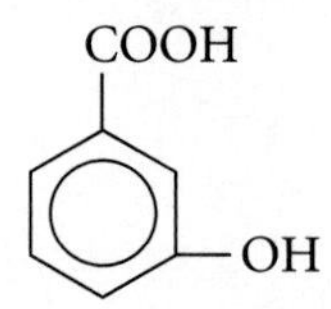

Type ____________________

amine and acid (amino acid)

273

$CH_3CH(NH_2)COOH$

Type ____________________

274

$$\underset{\overset{\|}{O}}{CH_3C}CH_2CH_2NH_2$$

Type ____________________

ketone and amine, or aminoketone

275

$CH_3CH_2CH(OH)COOH$

Type ____________________

alcohol and acid

ESTERS

A compound that contains a $-\overset{\overset{O}{\|}}{C}-O-C$ group is called an *ester*. The following compound is an ester because it contains such a group:

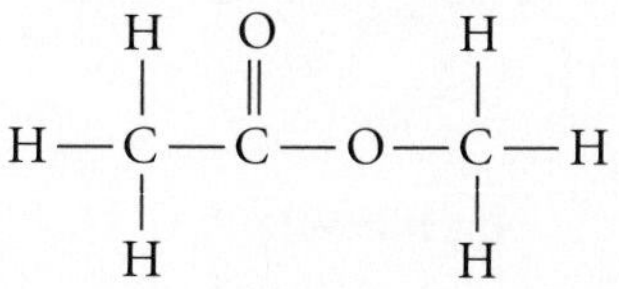

$$H-\underset{\underset{H}{|}}{\overset{\overset{H}{|}}{C}}-\overset{\overset{O}{\|}}{C}-O-\underset{\underset{H}{|}}{\overset{\overset{H}{|}}{C}}-H$$

276

Fats and oils are esters. They contain a ____________ group.

$-\overset{\overset{O}{\|}}{C}-O-C$

yes

277

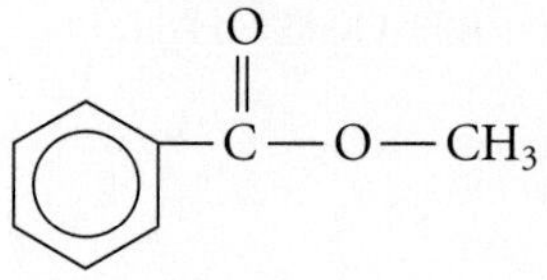

$$C_6H_5-\overset{\overset{\displaystyle O}{\|}}{C}-O-CH_3$$

Is this compound an ester? ____________

AMIDES

A compound that contains a $-\overset{\overset{\displaystyle O}{\|}}{C}-\underset{|}{N}H$ or $-\overset{\overset{\displaystyle O}{\|}}{C}-NH_2$ group is called an *amide.*

The following compound, niacinamide, which is derived from the B-vitamin niacin, is an example of an amide.

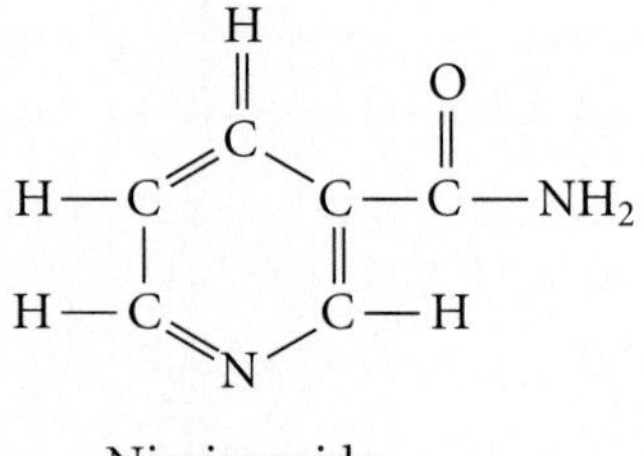

$$\begin{array}{ccccc} & & H & & O \\ & & \| & & \| \\ & & C & & \\ H-C & & & C & -C-NH_2 \\ | & & & \| & \\ H-C & & & C & -H \\ & & N & & \end{array}$$

Niacinamide

278

Cells link amino acids together. The bond holding amino acids together is called an *amide* or a *peptide bond.*

A dipeptide is formed by the linkage of how many amino acids?

____

The amino acids in a dipeptide are held together by an amide or a(n) ____________ bond.

A tripeptide consists of ____________ (how many?) amino acids held together by ____________ or ____________ bonds.

2; peptide; 3; amide; peptide

SULFHYDRYLS

A compound that contains an —SH group is called a *sulfhydryl* or a *thiol.*

The following compound, cysteine, one of the essential amino acids, is a sulfhydryl, or thiol, because it contains an —SH group.

$$\begin{array}{c} COOH \\ | \\ NH_2—C—H \\ | \\ CH_2—SH \end{array}$$

Cysteine

279

The activity of some enzymes is due to the presence of sulfhydryl groups. Cysteine also contains which other functional groups?

____

acid (or —COOH) and —NH_2 (or amine)

alcohol; sulfur; hydrogen

280

The prefix *thio* refers to the element sulfur, whereas the suffix *ol* indicates a(n) ____________. So, a thiol is a sulfur alcohol and is made from the elements ____________ and ____________.

DISULFIDES

Organic disulfides or disulfide bridges contain an —S—S— group.

The following compound, cystine, contains a disulfide bridge.

$$\begin{array}{ccccc} & COOH & & & COOH \\ & | & & & | \\ NH_2— & CH & & NH_2— & CH \\ & | & & & | \\ & CH_2 & —S—S— & & CH_2 \end{array}$$

Cystine

2; sulfur; 2; sulfur

The prefix *di* indicates the number ____________. Sulfide refers to the element ____________. So, a disulfide bridge contains ____________ (how many?) atoms of the element ____________.

282

Note the following abbreviated structure of human insulin.

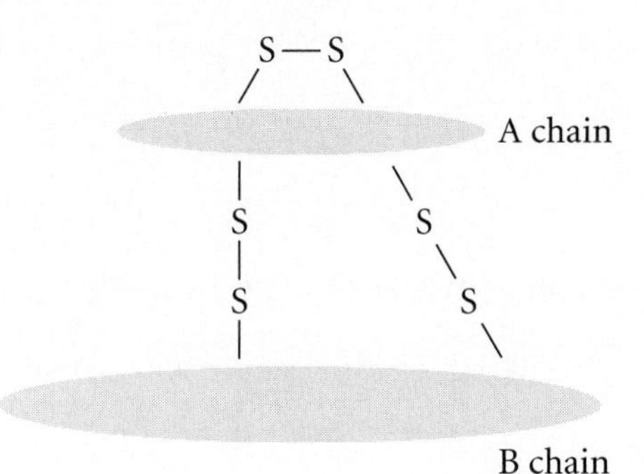

What type of bonds or linkages holds the A and B chains together?

____

How many bonds of this type are present in human insulin? __________

disulfide; 3

PHOSPHATES

A compound that contains a $—OPO_3H_2$ group is called a *phosphate*. This group may also be written as

$$-O-\overset{\overset{\displaystyle O}{\|}}{\underset{\underset{\displaystyle OH}{|}}{P}}-OH$$

The following compound is a phosphate because it contains the $—OPO_3H_2$ group:

$$H-\overset{\overset{\displaystyle H}{|}}{\underset{\underset{\displaystyle H}{|}}{C}}-\overset{\overset{\displaystyle O}{\|}}{C}-O-\overset{\overset{\displaystyle O}{\|}}{\underset{\underset{\displaystyle OH}{|}}{P}}-OH$$

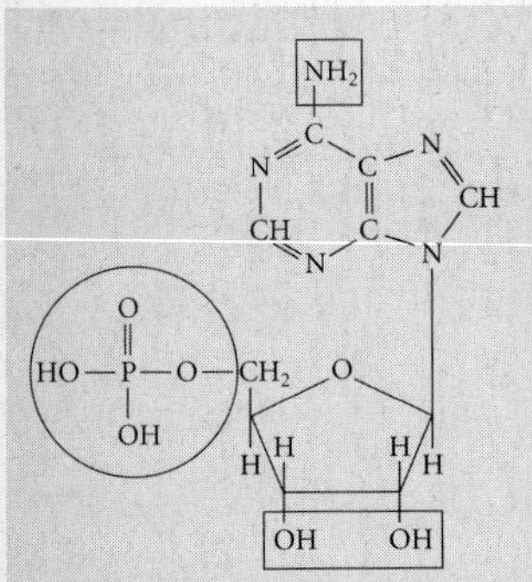

NH$_2$
N C C N CH
CH C N
N
O
HO—P—O—CH$_2$ O
OH
H H H H
OH OH

283

The structure below represents adenosine monophosphate (AMP). Draw a circle around the phosphate group. This compound also contains an amine and two alcohol groups. Draw a square around the amine group, and a rectangle around the OH groups.

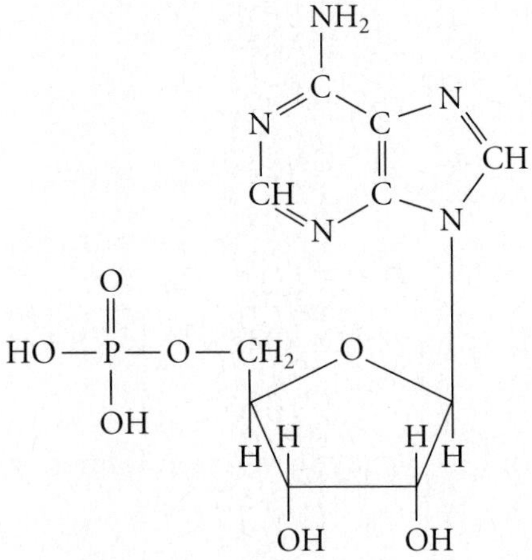

NH$_2$
C
N C N
CH
CH C N
N
O
HO—P—O—CH$_2$ O
OH
H H H H
OH OH

SUMMARY OF FUNCTIONAL GROUPS

Name the groups present in the compounds in problems 284–296.

284

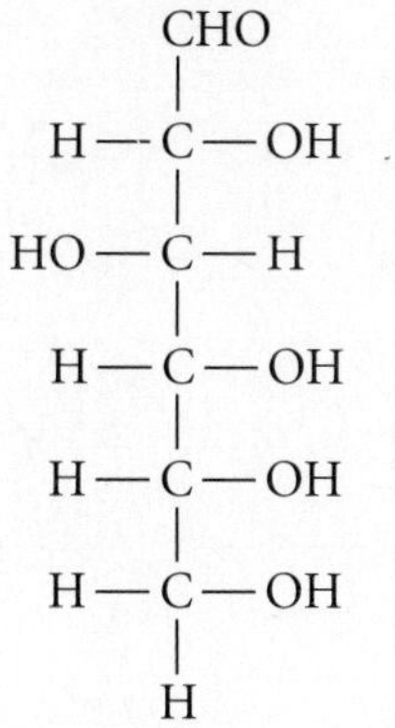

```
          CHO
           |
     H  —  C  —  OH
           |
    HO  —  C  —  H
           |
     H  —  C  —  OH
           |
     H  —  C  —  OH
           |
     H  —  C  —  OH
           |
           H
```

Glucose

aldehyde and alcohols (more than one alcohol group)

285

$CH_3CH(NH_2)COOH$

Alanine

amine and acid (amino acid)

286

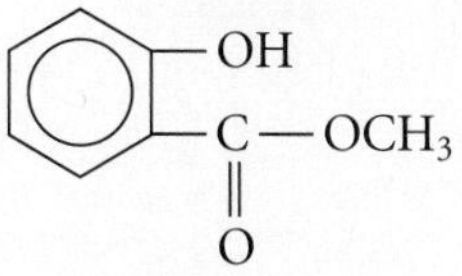

```
 [benzene ring] — OH
 [benzene ring] — C — OCH3
                  ||
                  O
```

Methyl salicylate

alcohol and ester

aldehyde and alcohols

287

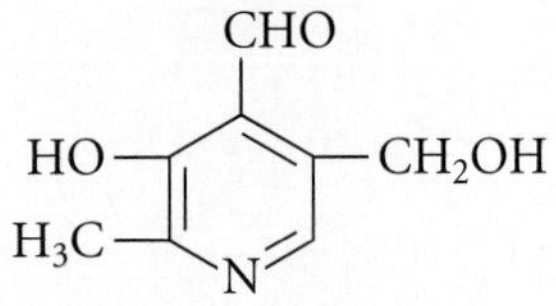

CHO
HO — — CH_2OH
H_3C — N

Pyridoxal
(one of the B-vitamins)

---

one phosphate group and two ester groups

288

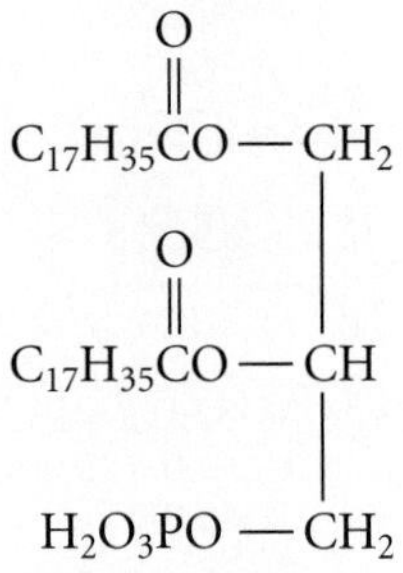

$C_{17}H_{35}C(=O)O - CH_2$
$C_{17}H_{35}C(=O)O - CH$
$H_2O_3PO - CH_2$

Lipid

---

two alcohol groups, a ketone group, and an acid group

289

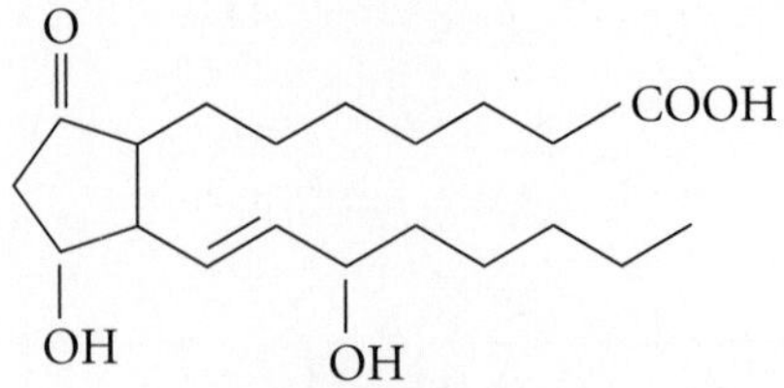

O, COOH, OH, OH

Prostaglandin

290

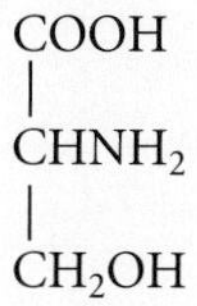

$$\begin{array}{l} COOH \\ | \\ CHNH_2 \\ | \\ CH_2OH \end{array}$$

Serine

an alcohol group, an amine group, and an acid group

291

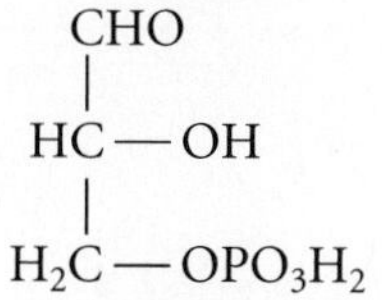

$$\begin{array}{l} \quad\;\; CHO \\ \quad\;\; | \\ \;\; HC-OH \\ \quad\;\; | \\ H_2C-OPO_3H_2 \end{array}$$

Glyceraldehyde-3-P

aldehyde, alcohol, and phosphate groups

292

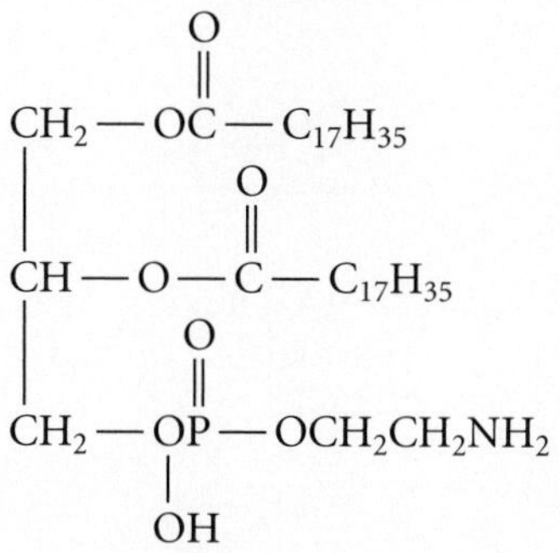

$$\begin{array}{l} CH_2-O\overset{\overset{O}{\|}}{C}-C_{17}H_{35} \\ | \\ CH-O-\overset{\overset{O}{\|}}{C}-C_{17}H_{35} \\ | \\ CH_2-O\overset{\overset{O}{\|}}{\underset{\underset{OH}{|}}{P}}-OCH_2CH_2NH_2 \end{array}$$

Cephalin

two ester groups, a phosphate group, and an amine group

an amide group

293

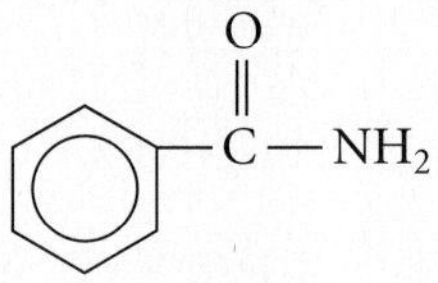

$$\text{C}_6\text{H}_5-\overset{\displaystyle O}{\overset{\|}{C}}-NH_2$$

---

a thiol (or sulfhydryl) group

294

$CH_3CH_2CH_2SH$

---

two acid groups, two amine groups, and one disulfide bridge

295

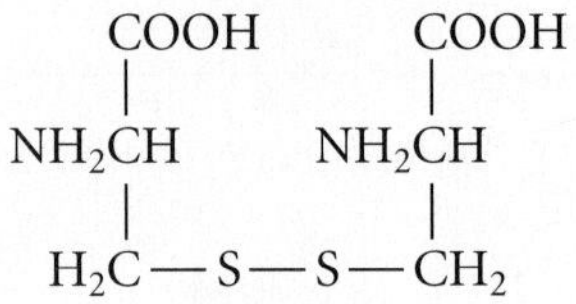

$$\begin{array}{ccccc} COOH & & & & COOH \\ | & & & & | \\ NH_2CH & & & & NH_2CH \\ | & & & & | \\ H_2C & -S- & & -S- & CH_2 \end{array}$$

---

an acid group, an amine group, and a sulfhydryl group

296

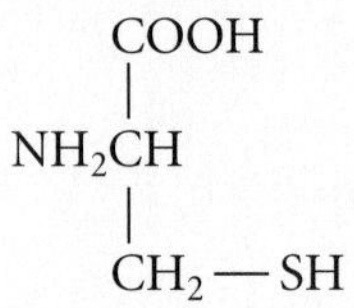

$$\begin{array}{l} COOH \\ | \\ NH_2CH \\ | \\ CH_2-SH \end{array}$$

297

The following diagram illustrates part of the shape of a protein molecule.

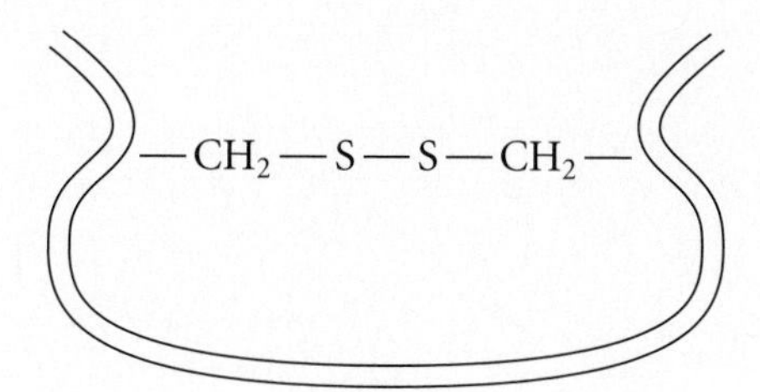

What type of bond or bridge determines the conformation of such a protein? ____________

disulfide

298

Polypeptides are made from many amino acids bonded together by ____________ or ____________ bonds.

amide; peptide

PART III
Biochemistry

Carbohydrates 10

Carbohydrates contain the elements carbon, hydrogen, and oxygen, and only those elements. A compound containing elements other than these three is not a carbohydrate.

The word *hydrate* refers to water. Many carbohydrates contain hydrogen and oxygen in the same ratio as in water—2:1.

299

Which of the following compounds are carbohydrates?

__ C_6H_6 __ $C_{12}H_{22}O_{11}$ __ $C_5H_{10}O_5$
__ $C_6H_{12}O_6N$ __ C_2H_6O

$C_{12}H_{22}O_{11}$ and $C_5H_{10}O_5$

300

a. In the list of compounds in problem 299, why is $C_6H_{12}O_6N$ not a carbohydrate? ______________________________

b. Why are C_6H_6 and C_2H_6O not carbohydrates? ______________________________

a. $C_6H_{12}O_6N$ contains nitrogen, and carbohydrates contain only C, H, and O
b. C_6H_6 contains no oxygen, and C_2H_6O has the wrong ratio of H to O

a. one
b. one

301

Carbohydrates are divided into three types. One type of carbohydrate is the *monosaccharide*.

a. The word *saccharide* means *simple sugar*. The prefix *mono* means ____________.

b. Therefore, monosaccharides are simple sugars, each containing only ____________ simple sugar.

all of them

302

Monosaccharides and disaccharides have names ending in *ose*. Which of the following are monosaccharides or disaccharides?

__ glucose __ fructose __ galactose
__ arabinose __ erythrose

5

303

Monosaccharides are named according to the number of carbon atoms they contain. A pentose contains ____________ carbon atoms.

An example of a pentose is ribose, $C_5H_{10}O_5$. The structure of ribose is:

```
CH2OH   O        OH
 |    /    \      |
 C   H      H    C
 |  \|      |  / |
 H   C======C    H
     |      |
     OH     OH
```

In this type of structure the ring, consisting of carbon atoms at each corner and an oxygen atom where indicated, is considered to be in a plane perpendicular to the paper. The heavy line in the ring indicates the section closest to you; the light lines, the part farther from you. The —H and —OH groups are above and below the plane as indicated.

Ribose occurs in ribonucleic acid, RNA, which will be discussed later.

a. one less oxygen
b. one less oxygen

304

Compare the structures of ribose, $C_5H_{10}O_5$, and deoxyribose, $C_5H_{10}O_4$.

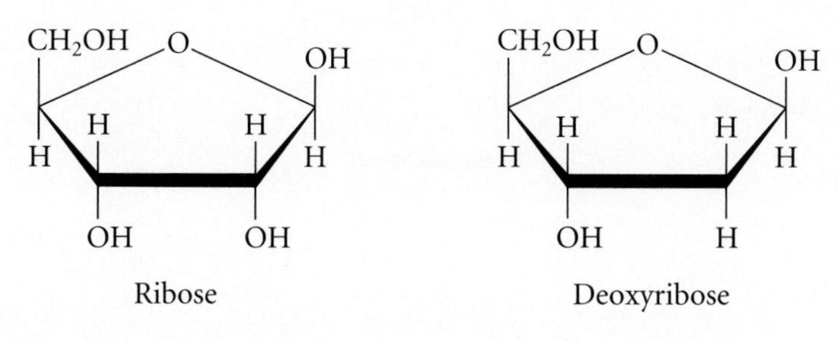

a. The prefix *deoxy* means ____________.

b. Deoxyribose contains ____________ than does ribose.

DNA

305

Ribose occurs in ribonucleic acid, RNA. Deoxyribose occurs in deoxyribonucleic acid, ____________.

a. 5
b. 6

306

a. A pentose is a monosaccharide containing ____________ carbon atoms.

b. A hexose is a monosaccharide containing ____________ carbon atoms.

A primary hexose in the body is glucose, $C_6H_{12}O_6$. The oxidation of glucose supplies much of the energy the body needs. The structure of glucose is:

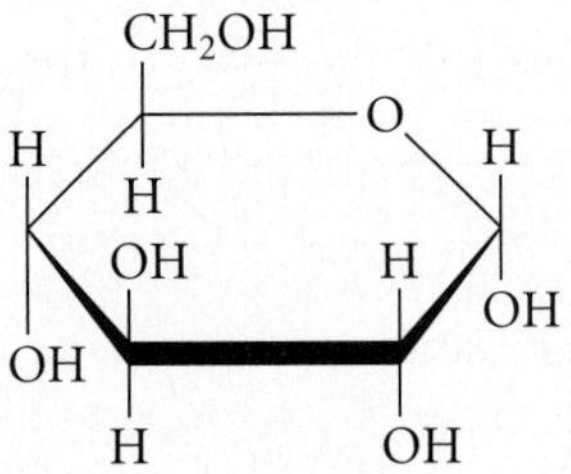

CH_2OH
O
H H
H
OH H
OH
OH
H OH

ose

307

The second type of carbohydrate is the *disaccharide*. Both monosaccharides and disaccharides have names ending in ____________.

308

The prefix *di* means *two*. Disaccharides are formed by the combination of two monosaccharides, with water also being produced. An enzyme is required for this reaction.

$$\text{monosaccharide} + \text{monosaccharide} \xrightarrow{\text{enzyme}} \text{disaccharide} + \text{water}$$

Conversely, disaccharides react with water to produce two monosaccharides.

$$\text{disaccharide} + \text{water} \xrightarrow{\text{enzyme}} \text{monosaccharide} + \_\_\_\_\_\_\_\_\_\_$$

These reactions may be written as shown on the following page.

The bond holding the two halves of the disaccharide together is called a *glycosidic bond*, or a *glycosidic linkage*.

monosaccharide

309

When a disaccharide reacts with water, two monosaccharides are produced. Such a reaction is called *hydrolysis*, which is defined as the breaking apart of a molecule by reaction with water.

Digestion involves hydrolysis. During digestion, molecules are

__ broken down into smaller molecules
__ built up into larger molecules

broken down into smaller molecules

monosaccharides

310

Disaccharides that have undergone hydrolysis yield two ____________.

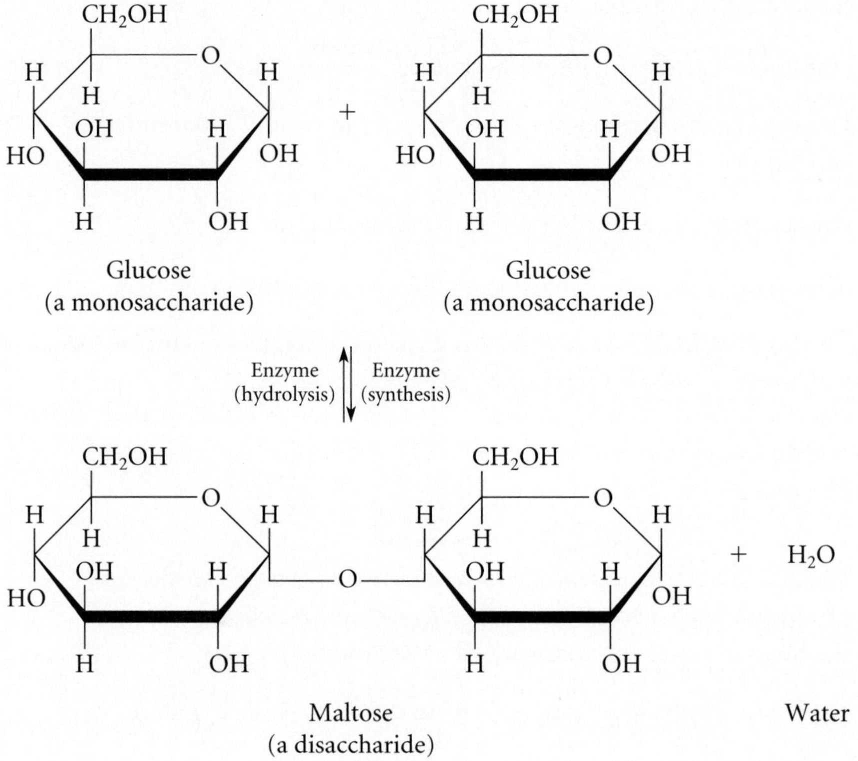

CH_2OH H H OH H HO OH H OH — Glucose (a monosaccharide) + Glucose (a monosaccharide)

Enzyme (hydrolysis) ⇅ Enzyme (synthesis)

Maltose (a disaccharide) + H_2O Water

one; two

311

Monosaccharides are made up of ____________ simple sugar(s).
Disaccharides are made up of ____________ simple sugar(s).

312

The hydrolysis of a disaccharide may be written as:

$$\underset{\text{disaccharide}}{C_{12}H_{22}O_{11}} + \underset{\text{water}}{H_2O} \xrightarrow{\text{enzyme}} \underset{\text{monosaccharide}}{C_6H_{12}O_6} + \underset{\text{monosaccharide}}{C_6H_{12}O_6}$$

Sucrose, $C_{12}H_{22}O_{11}$, is a disaccharide. The hydrolysis of sucrose yields the two ____________, glucose and fructose.

monosaccharides

313

There are three common disaccharides: *sucrose*, *maltose*, and *lactose*. These disaccharides all have the same molecular formula, $C_{12}H_{22}O_{11}$. Lactose, also called milk sugar, is found in milk. Maltose, or malt sugar, is found in sprouting grain. Sucrose, or cane sugar, is found in ____________.

sugarcane

314

The third type of carbohydrate is the *polysaccharide*. The prefix *poly* means *many*, so polysaccharides, on hydrolysis, yield many __________.

monosaccharides

315

Recall that hydrolysis means reaction with ____________ to produce ____________ (smaller/larger) molecules.

water; smaller

$C_{12}H_{22}O_{11}$

316

Starch is a polysaccharide. Its molecular formula is $(C_6H_{10}O_5)_n$, where *n* is some large number. Upon complete hydrolysis, starch yields many monosaccharides:

$$\underset{\text{Starch, a polysaccharide}}{(C_6H_{10}O_5)_n} + nH_2O \xrightarrow{\text{enzyme}} \underset{\text{Monosaccharides}}{nC_6H_{12}O_6}$$

Which of the following formulas represents a disaccharide?

__ $C_5H_{10}O_5$ __ $C_{12}H_{22}O_{11}$
__ $(C_6H_{10}O_5)_n$ __ $C_6H_{12}O_6$

$(C_6H_{10}O_5)_n$; $C_5H_{10}O_5$ and $C_6H_{12}O_6$

317

Which of the formulas listed in problem 316 represents a polysaccharide? ____________

Which are monosaccharides? ____________

a. glucose, fructose, galactose
b. lactose, maltose, sucrose
c. starch, cellulose, glycogen

318

Other examples of polysaccharides are *cellulose*, found in plants, and *glycogen*, found in animals. Plants use the polysaccharide cellulose primarily for support. Animals store carbohydrate in the form of a polysaccharide, glycogen.

a. Name three monosaccharides: ____________

b. Name three common disaccharides: ____________

c. Name three polysaccharides: ____________

319

Disaccharides are held together by a glycosidic bond. Polysaccharides are held together by ____________ bonds.

many glycosidic

320

a. Hydrolysis of a disaccharide yields ___________________________.

b. Hydrolysis of a polysaccharide yields ___________________________.

a. two monosaccharides
b. many monosaccharides

PROPERTIES OF CARBOHYDRATES

Monosaccharides are crystalline compounds, soluble in water and sweet to the taste, and need no digestion in order to be absorbed into the blood stream.

Disaccharides are crystalline, water-soluble, and sweet to the taste and must be hydrolyzed (digested) to monosaccharides before they can be absorbed and used by the body for energy.

Polysaccharides are not water-soluble and are not crystalline. They form colloidal dispersions instead of solutions. They are not sweet and must be hydrolyzed before being absorbed. (Cellulose is not hydrolyzed by enzymes in the human gastrointestinal tract.)

a. **mono- and disaccharides**
b. **polysaccharides**
c. **mono- and disaccharides**
d. **polysaccharides**
e. **mono- and disaccharides**
f. **polysaccharides**
g. **monosaccharides**

321

a. Which type(s) of carbohydrate is(are) soluble in water? ____________

b. Which type(s) of carbohydrate form(s) a colloidal suspension in water? ____________

c. Which type(s) of carbohydrate is(are) crystalline? ____________

d. Which is(are) not crystalline? ____________

e. Which is(are) sweet? ____________

f. Which is(are) not sweet? ____________

g. Which require(s) no digestion? ____________

Lipids 11

Lipids include such substances as fats, phospholipids, and steroids. Lipids are insoluble in water. Lipids contain the same elements as carbohydrates, but in lipids the proportion of oxygen is much lower than in carbohydrates.

322

Lipids contain the elements ____________, ____________, and ____________.

carbon; hydrogen; oxygen

FATS

323

Fats are esters. Esters are formed by the reaction of an organic acid with an alcohol. In fats, the organic acid is called a *fatty acid* and the alcohol is *glycerol* (sometimes called glycerin). The reaction is:

$$\text{fatty acids} + \text{glycerol} \xrightarrow{\text{enzymes}} \text{fat} + \text{water}$$

The reverse reaction is the hydrolysis of a fat.

Thus, the hydrolysis of a fat yields ____________ + ____________.

fatty acids; glycerol

fat

fatty acid; glycerol; esters

324

Fatty acids and glycerol are products of the hydrolysis of a ___________.

325

The chemical equation for the hydrolysis of a fat may be written as

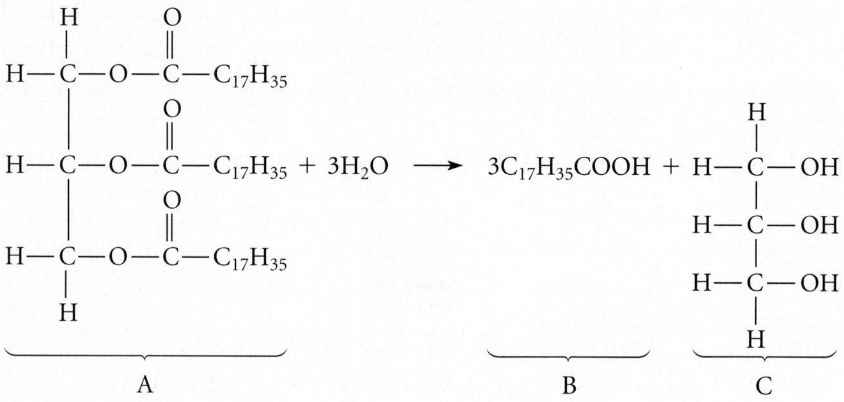

$$\underbrace{\begin{array}{l} \mathrm{H-\overset{H}{\underset{|}{\overset{|}{C}}}-O-\overset{O}{\overset{\|}{C}}-C_{17}H_{35}} \\ \mathrm{H-\underset{|}{\overset{|}{C}}-O-\overset{O}{\overset{\|}{C}}-C_{17}H_{35}} \\ \mathrm{H-\underset{H}{\overset{|}{\underset{|}{C}}}-O-\overset{O}{\overset{\|}{C}}-C_{17}H_{35}} \end{array}}_{A} + 3H_2O \longrightarrow \underbrace{3C_{17}H_{35}COOH}_{B} + \underbrace{\begin{array}{l} \mathrm{H-\overset{H}{\overset{|}{C}}-OH} \\ \mathrm{H-\underset{|}{\overset{|}{C}}-OH} \\ \mathrm{H-\underset{H}{\underset{|}{C}}-OH} \end{array}}_{C}$$

Substance A is a fat.

Substance B is a ___________.

Substance C is an alcohol called ___________.

Substance A contains what type of groups? ___________

326

a. How do you know that substance B is an acid? ________________

b. How do you know that substance C is an alcohol? ________________

c. Why is substance A called an ester? ________________

a. It contains —COOH groups.
b. It contains —OH groups.
c. It contains

$$\overset{\overset{\displaystyle O}{\|}}{C}-O-C \quad \left(\overset{\overset{\displaystyle O}{\|}}{C}-O-C\right)$$

groups.

327

a. The hydrolysis of a polysaccharide yields ________________.

b. The hydrolysis of a disaccharide yields ________________.

c. The hydrolysis of a fat yields ________________.

a. many monosaccharides
b. two monosaccharides
c. fatty acids and glycerol

328

Fatty acids that contain double bonds are called *unsaturated.* Unsaturated fatty acids tend to be more liquid than saturated fatty acids. A liquid fat is called an *oil.*

Increasing the number of double bonds between carbon atoms in fatty acids of a fat molecule should make the fat more ____________ (solid/liquid). Decreasing the number of double bonds in a fat will make the fat more ____________ (solid/liquid).

liquid; solid

329

Which are unsaturated, fats or oils? ____________

oils

those containing many double bonds

330

Unsaturated compounds contain double bonds. What kind of compounds would be called polyunsaturated? ______________________

solid

331

Three fatty acids, butyric (C_3H_7COOH), palmitic ($C_{15}H_{31}COOH$), and stearic ($C_{17}H_{35}COOH$) acid, contain only single bonds. When they are present in a lipid, will that lipid tend to become solid or liquid? ____________

some double bonds

332

Arachidonic acid is unsaturated. This means that it contains __________ (only single bonds/some double bonds) in its structure.

an oil; the fatty acid is unsaturated

333

A lipid that undergoes hydrolysis yields arachidonic acid and glycerol. Was the lipid a fat or an oil? ____________

How do you know? ____

nonpolar

334

Lipids are insoluble in water. They are soluble in organic solvents such as gasoline, ether, carbon tetrachloride, and benzene. Organic solvents are nonpolar. Recall that, in general, polar compounds dissolve in polar liquids, and nonpolar compounds dissolve in nonpolar liquids. Are lipids polar or nonpolar? ____________

PHOSPHOLIPIDS

335

The hydrolysis of a phospholipid may be written as:

$$\text{phospholipid} \xrightarrow[\text{enzymes}]{\text{hydrolysis}} \text{fatty acids} + \text{glycerol} + \text{phosphoric acid} + \text{an N-compound}$$

The hydrolysis of a fat yields ____________ + ____________.

fatty acids; glycerol

336

The N-compounds in phospholipids are a certain group of nitrogen-containing compounds. Among them are serine, choline, inositol, and ethanolamine. Which is present determines the type of phospholipid, as will be discussed in the following problems.

The hydrolysis of a phospholipid yields ______________________________ ______________________________.

fatty acids, glycerol, phosphoric acid, and a nitrogen compound

337

a. Lipids contain which elements? ____________

b. Phospholipids contain which elements? ____________

a. C, H, O
b. C, H, O, P, N

338

Lecithin is a phospholipid in which the N-compound is *choline*. Lecithin is a main component of cell membranes and is involved in the transportation of fats in the body. The hydrolysis of lecithin yields fatty acids, glycerol, phosphoric acid, and ____________.

choline

fatty acids; glycerol; phosphoric acid; ethanolamine

339

Cephalin is a phospholipid involved in the blood-clotting process. In cephalin the N-compound is ethanolamine.

$$\text{cephalin} \xrightarrow[\text{enzymes}]{\text{hydrolysis}}$$ ____________, ____________, ____________, and ____________.

STEROIDS

a. insoluble
b. insoluble
c. soluble

340

Another group, called steroids, is classified along with lipids because of similar solubility characteristics.

a. Lipids are ____________ (soluble/insoluble) in water.

b. Steroids are ____________ (soluble/insoluble) in water.

c. Steroids are ____________ (soluble/insoluble) in organic liquids.

a. no
b. no

341

Steroids are high-molecular-weight tetracyclic (4-ring) compounds whose structure is similar to this:

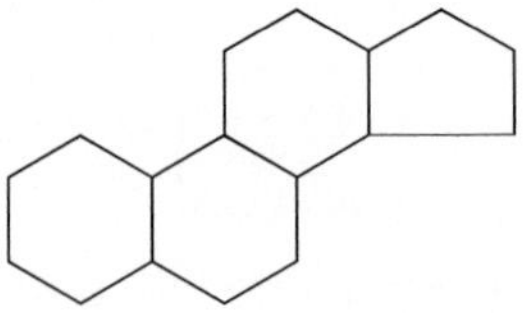

a. Do steroids contain fatty acids? ____________

b. Do steroids contain glycerol? ____________

342

Examples of steroids in the body are cholesterol, bile salts, sex hormones, and hormones of the adrenal cortex.

a. Does cholesterol have a steroid structure? ____________

b. The ending *ol* in cholesterol indicates that it is what type of compound? ____________

a. yes
b. an alcohol

Proteins 12

FORMATION AND HYDROLYSIS

343

Proteins contain the same elements as do carbohydrates and fats, except that proteins always contain one additional element: nitrogen.

All proteins contain the four elements: ______________________.

carbon, hydrogen, oxygen, nitrogen (in any order)

344

Some proteins also contain additional elements, such as sulfur, phosphorus, or iron. The hydrolysis of proteins yields amino acids. Likewise, the combination of amino acids yields proteins.

$$\text{protein} \underset{\text{combination}}{\overset{\text{hydrolysis}}{\rightleftharpoons}} \text{amino acids}$$

Amino acids contain what functional groups? ______________________

amine and organic acid, or carboxyl (or $-NH_2$ and $-COOH$)

—COOH or carboxyl

345

The structure of an amino acid can be represented as

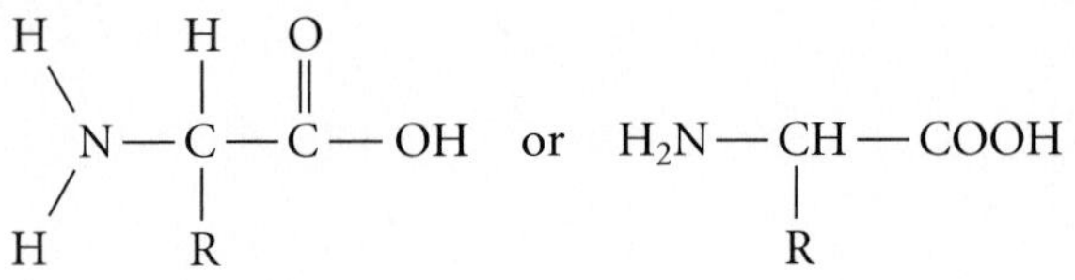

$$\begin{array}{c} H \quad\; H \;\; O \\ \diagdown \quad\; | \quad \| \\ N - C - C - OH \\ \diagup \quad\; | \\ H \quad\; R \end{array} \quad \text{or} \quad \begin{array}{c} H_2N - CH - COOH \\ | \\ R \end{array}$$

where the "R" represents either a hydrogen or carbon-hydrogen groups.

Amino acids (and also proteins) can act as buffers because the $—NH_2$ group acts as a H^+ acceptor and the ____________ group acts as a H^+ donor.

a. 3
b. many (or more than 3)

346

When two amino acids combine, the product is called a *dipeptide.*

a. A *tripeptide* would be formed when ____________ amino acids combine.

b. A *polypeptide* would be formed when ____________ amino acids combine.

amino acids

347

Note the formulas for alanine and glycine:

$$\begin{array}{c} CH_3CHCOOH \\ | \\ NH_2 \\ \text{Alanine} \end{array} \qquad \begin{array}{c} CH_2COOH \\ | \\ NH_2 \\ \text{Glycine} \end{array}$$

Both would be classified as what type of compound? ____________

348

When these two amino acids react, the following reactions are possible:

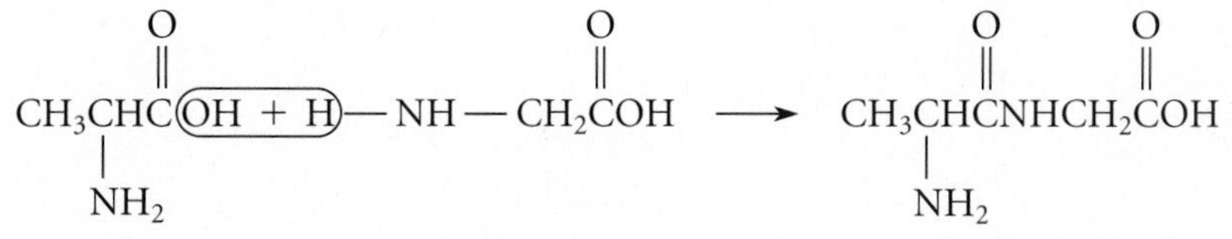

$$CH_3CH(NH_2)C(=O)OH + H\text{—}NH\text{—}CH_2C(=O)OH \longrightarrow CH_3CH(NH_2)C(=O)NHCH_2C(=O)OH$$

Alanine Glycine Alanyl-glycine (ala-gly)

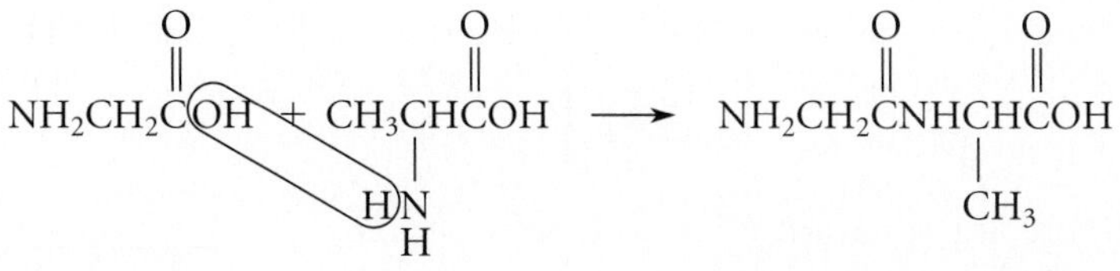

$$NH_2CH_2C(=O)OH + CH_3CH(NHH)C(=O)OH \longrightarrow NH_2CH_2C(=O)NHCH(CH_3)C(=O)OH$$

Glycine Alanine Glycyl-alanine (gly-ala)

Note that in both of these equations, the reaction is between the OH of a(n) ____________ group and the H of a(n) ____________ group.

organic acid (or carboxyl); amine

349

When glycine and alanine react, what type of compound is formed?

____

a dipeptide

350

a. The abbreviation *ala* represents ____________.

b. The abbreviation *ala-gly* represents ____________.

a. alanine
b. alanyl-glycine

a tripeptide

351

If the amino acids glycine, alanine, and valine were combined in one compound, what type would it be? ____________

ala-gly-val

352

What would be the abbreviation for a tripeptide containing alanine, glycine, and valine, in that order? ____________

a tripeptide containing glycine, valine, and alanine, combined in that order

353

What would the abbreviation *gly-val-ala* represent? ____

many; peptide

354

A dipeptide consists of two amino acids held together by a *peptide bond*. A tripeptide consists of three amino acids held together by peptide bonds. A polypeptide consists of ____________ amino acids held together by ____________ bonds.

amide

355

Peptide bonds are also called ____________ bonds. (See problem 278.)

356

In a peptide bond (see problem 348), the bond occurs between a ____________ atom of one amino acid and a ____________ atom of another amino acid.

C; N

STRUCTURE

The *primary structure* of a protein refers to the linear sequence of amino acids in that protein. An example of a primary structure is:

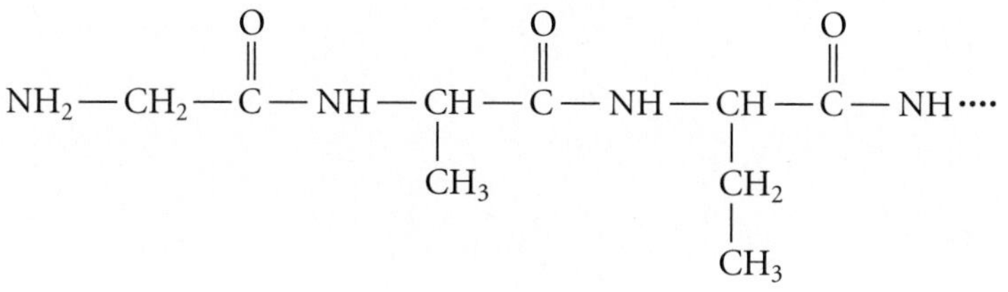

$$
\begin{array}{ccccccccccccccccc}
 & & & & O & & & & O & & & & O & & \\
 & & & & \| & & & & \| & & & & \| & & \\
NH_2 & - & CH_2 & - & C & - & NH & - & CH & - & C & - & NH & - & CH & - & C - NH\cdots \\
 & & & & & & & & | & & & & & & | & & \\
 & & & & & & & & CH_3 & & & & & & CH_2 & & \\
 & & & & & & & & & & & & & & | & & \\
 & & & & & & & & & & & & & & CH_3 & &
\end{array}
$$

Amino acid Amino acid Amino acid

357

What type of bond holds the amino acids together in the primary structure? ____________

Note that all proteins have an amino ($—NH_2$) terminus and a carboxyl ($—COOH$) terminus.

peptide or amide

The *secondary structure* of a protein refers to the regular recurring arrangement of the amino acid chain (its primary structure). One such arrangement, called the α-helix, occurs when the amino acid chain forms a spiral or coil, as shown here:

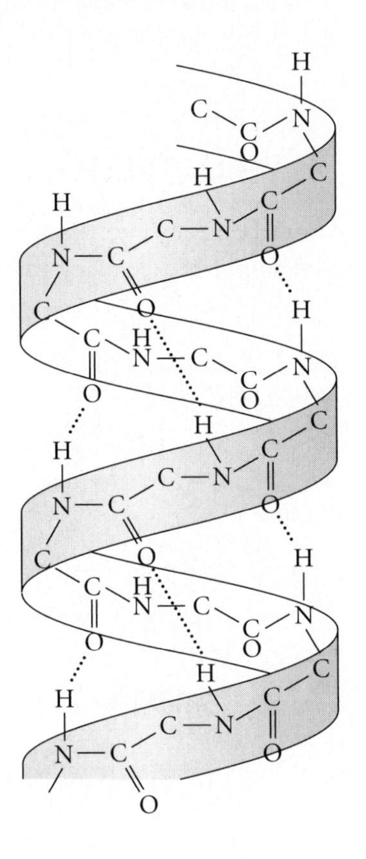

hydrogen

358

The coil consists of loops of the linear arrangement of the amino acids. These loops are held together by bonds between the O of the C=O of one amino acid and the H of the NH of another amino acid.

Such a bond is called a ____________ bond (refer to problems 225–236 if you don't know).

359

Another type of secondary structure, called the β-pleated sheet, consists of parallel strands of polypeptides as shown in the following figure:

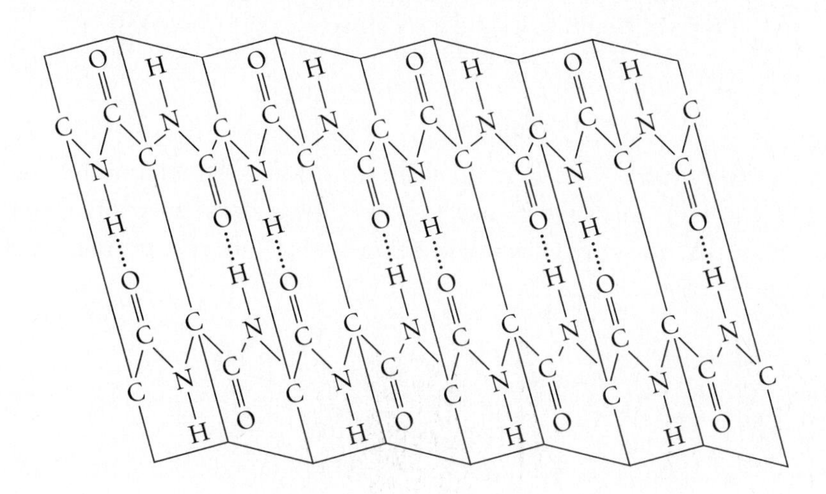

What type of bonds holds these sheets together? ___________

hydrogen

360

a. The linear sequence of amino acids in a protein is called the ___________ structure of that protein.

b. The regular recurring arrangement of the amino acid chains is called the ___________ structure of a protein.

a. primary
b. secondary

a. α-helix; β-pleated sheet
b. hydrogen

361

a. Two types of secondary structures are possible. They are the _____________ and the _____________.

b. What type of bond holds these secondary structures together? _____________

The *tertiary structure* of a protein refers to the specific folding and bending of the coils into layers or fibers, as shown here. It is this tertiary structure that gives proteins (and enzymes, which are also proteins) their biological activity.

Some proteins have a *quaternary structure* that occurs when several protein units, each with its own primary, secondary, and tertiary structure, combine to form a more complex unit, as shown in Figure **e**. An example of a protein with a quaternary structure is hemoglobin.

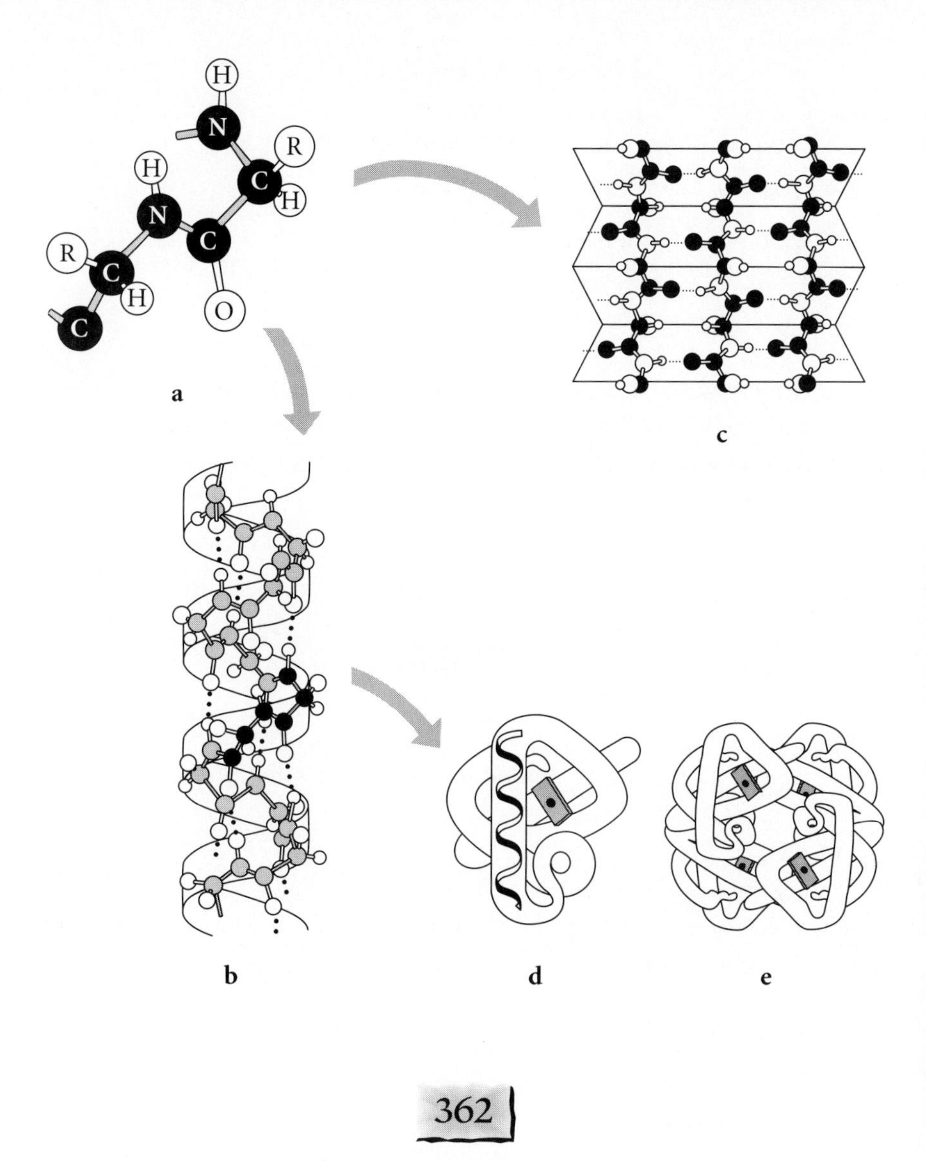

362

Look at the preceding five structures and label each as to primary, secondary, tertiary, or quaternary structure.

Which structure represents an α-helix? ____________

Which structure represents a β-pleated sheet? ____________

What type of bond holds the primary structures together? ____________

What type of bond holds the secondary structure together? ____________

a = primary;
b = secondary;
c = secondary;
d = tertiary;
e = quaternary

b; c; peptide; hydrogen

DENATURATION

weak

363

Hydrogen bonds are ____________ (strong/weak) bonds.

no

364

Therefore, hydrogen bonds are easily broken. If the hydrogen bonds in a protein are broken, can the protein easily maintain its structural shape?

____

cannot

365

If the hydrogen bonds in a protein are broken, the shape of the protein can change, making it incapable of performing its physiologic function. The protein is said to be *denatured*.

The activity of many proteins (and enzymes) depends upon the presence of an active site that "fits" into a specific substrate. In Figure **a** following, the active site is shown in the center of the structure.

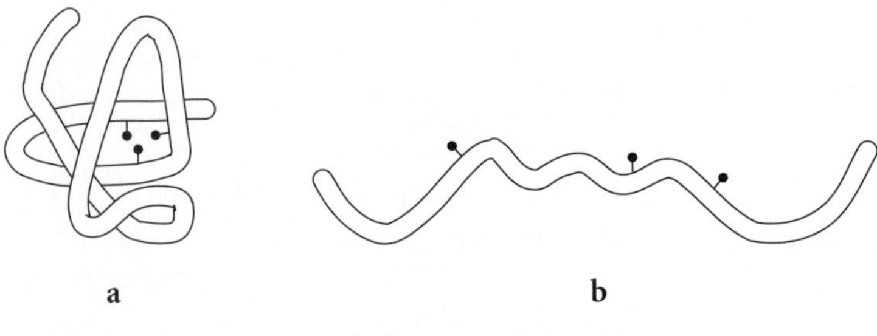

Enzymes

If the protein is denatured, the parts of the active site are no longer in close proximity (Figure **b**), can no longer "fit" into the substrate, and therefore ____________ (can/cannot) react.

366

If the hydrogen bonds in a protein can easily re-form, then the denaturation is said to be *reversible*. If the hydrogen bonds cannot re-form, then the process is termed *irreversible denaturation.*

a. In the denaturation of protein, ____________ bonds are broken.

b. If the broken hydrogen bonds can easily re-form, the process is called ____________ denaturation.

c. If the broken hydrogen bonds cannot re-form, the process is called ____________.

Factors that can denature proteins include strong acids, heat, alcohol, and salts of heavy metals.

a. hydrogen
b. reversible
c. irreversible denaturation

Nucleotides 13

ADENOSINE TRIPHOSPHATE

367

1

Nucleotides are formed by the reaction of a pentose (labeled Ⓢ for sugar), phosphoric acid (labeled Ⓟ), and a nitrogen-containing base (labeled Ⓝ), as shown:

Ⓝ — Ⓢ — Ⓟ

One example of a nucleotide is adenosine monophosphate, AMP, which is formed by the reaction of adenine (a nitrogen-containing base), ribose (a pentose), and phosphoric acid. From the name *adenosine monophosphate*, how many phosphoric acid molecules are present? ____________

368

The structure of adenosine monophosphate, AMP, is shown here. It can be abbreviated as adenosine-P.

ribose; adenine

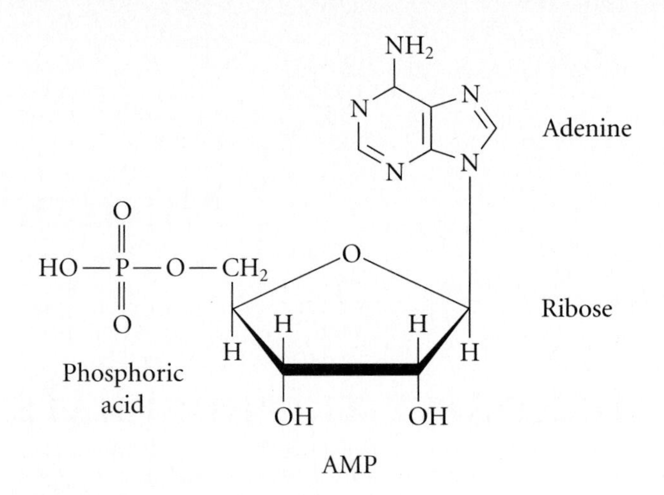

AMP contains which pentose? ____________

Which nitrogen-containing base? ____________

3

369

If a second molecule of phosphoric acid is bonded to the first phosphoric acid group, adenosine diphosphate, ADP, is formed. In adenosine triphosphate, ATP, how many phosphoric acid groups are present? ____________

energy

370

ATP provides a form of chemical energy that is usable by all body cells. The structure of ATP may be abbreviated as

adenosine — Ⓟ ~ Ⓟ ~ Ⓟ

Note that ATP contains two high-energy phosphate bonds indicated as ~.

In the reaction

ATP ⟶ ADP + Ⓟ + ?, or

adenosine — Ⓟ ~ Ⓟ ~ Ⓟ ⟶ adenosine — Ⓟ ~ Ⓟ + Ⓟ + ?,
what is produced other than ADP and P? ____________

371

When ADP is converted into AMP, is energy released? ____________

yes

372

To convert AMP back to ATP, what must be added? ____________

energy and two phosphoric acid groups

373

Much of the energy required to form ATP comes from the metabolism of glucose.

a. AMP contains __________ phosphoric acid groups and __________ high-energy phosphate bonds.

b. ADP contains __________ phosphoric acid groups and __________ high-energy phosphate bonds.

c. ATP contains __________ phosphoric acid groups and __________ high-energy phosphate bonds.

a. 1; 0
b. 2; 1
c. 3; 2

374

If two nucleotides are joined together, a dinucleotide is formed. One such dinucleotide is nicotinamide adenine dinucleotide, NAD.

a. Both nucleotides that make up NAD must contain which pentose? ____________

b. One nucleotide contains nicotinamide as the nitrogen-containing base; the other contains ____________ as the nitrogen-containing base.

a. ribose
b. adenine

flavin, adenine

375

Which nitrogen-containing bases are present in flavin adenine dinucleotide, FAD? ____________

Both dinucleotides, FAD and NAD, are involved in oxidation-reduction reactions in cells.

NUCLEIC ACIDS: DNA AND RNA

Nucleic acids are polymers of nucleotides. Consider the following structure of deoxyribonucleic acid, DNA, a nucleic acid. Note that each boxed segment is a nucleotide.

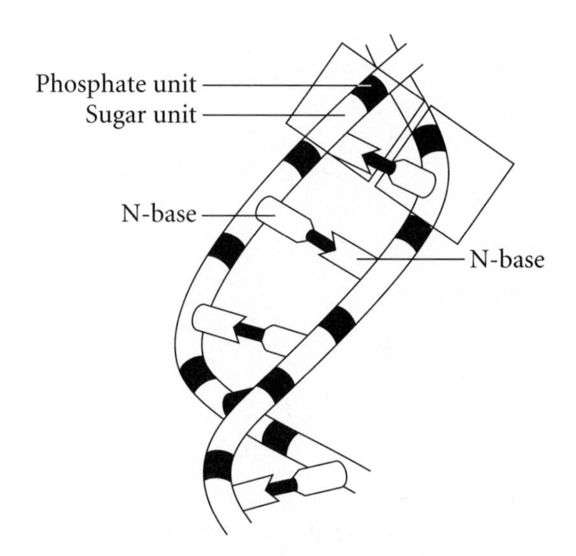

DNA consists of a double coil with the opposite sides held together by hydrogen bonds. The backbones of the coils are alternating sugar and phosphate groups, with the hydrogen bonding between the nitrogen-containing bases holding the sides together.

deoxyribose

376

In deoxyribonucleic acid, which pentose is a part of the coil backbone?

____

377

In DNA there are four nitrogen-containing bases. They are:

adenine, A
thymine, T
cytosine, C
guanine, G

In DNA adenine is always hydrogen-bonded to thymine, and cytosine is always hydrogen-bonded to guanine. Thus the structure of DNA may be represented as:

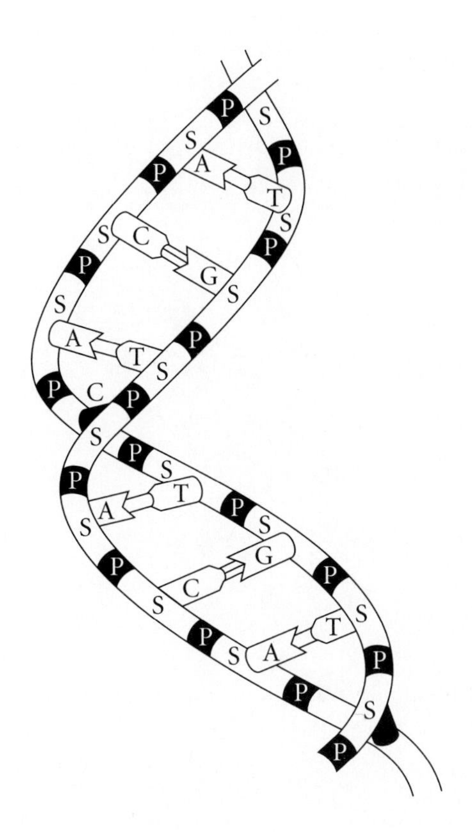

a. How many hydrogen bonds hold A and T together? ____________

b. How many hydrogen bonds hold C and G together? ____________

a. 2
b. 3

378

The four nitrogen-containing bases in DNA are: ______________________ ____.

adenine, A; thymine, T; guanine, G; and cytosine, C

379

In DNA, adenine is always hydrogen-bonded to ______________, and guanine is always bonded to ______________.

thymine; cytosine

380

Complete the structure of the following figure, indicating missing nitrogen-containing bases and hydrogen bonds.

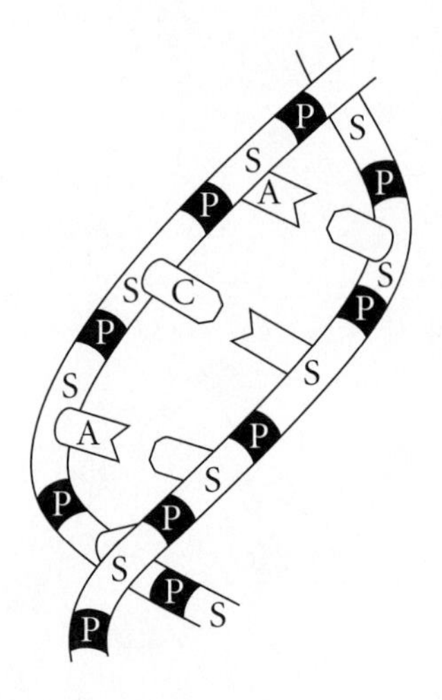

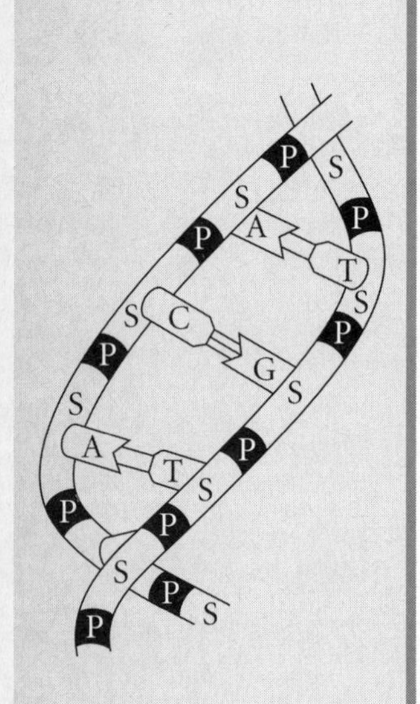

381

A second nucleic acid is ribonucleic acid, RNA. In RNA the pentose is ribose; in DNA the pentose is ____________.

deoxyribose

382

In RNA, A is always bonded to uracil, U. In DNA, A is always bonded to ____________.

T

383

a. In DNA, C bonds to ____________ with ____________ (how many?) hydrogen bonds?

b. In RNA, C bonds to ____________ with ____________ (how many?) hydrogen bonds?

c. In DNA, A bonds to ____________ with ____________ (how many?) hydrogen bonds?

d. In RNA, A bonds to ____________ with ____________ (how many?) hydrogen bonds?

e. In DNA the sugar is ____________.

f. In RNA the sugar is ____________.

a. G; 3
b. G; 3
c. T; 2
d. U; 2
e. deoxyribose
f. ribose

Enzymes 14

Enzymes are proteins that act as biological catalysts. A catalyst increases the rate of a reaction but is not changed itself.

Pepsin is an enzyme found in gastric juice. It increases the rate of reaction for the digestion of protein in the stomach. Note that pepsin is an enzyme for a particular substance: protein. In general, enzymes are highly specific; catalysts are nonspecific.

384

Enzymes are ____________ (specific/nonspecific) catalysts.

specific

385

a. What effect does an enzyme have on the rate of a reaction? ____________

b. Is an enzyme changed by the reaction? ____________

a. it increases it
b. no

NAMES OF ENZYMES

a. carbohydrates
b. lipids

386

Most enzymes have names ending in *ase*. Some enzymes are named for the type of reaction they catalyze. *Hydrolases* are enzymes that catalyze hydrolysis reactions. *Oxidases* are enzymes that catalyze oxidation reactions.

a. Carbohydrases are enzymes that catalyze the hydrolysis of ____________.

b. Lipases are enzymes that catalyze the hydrolysis of ____________.

a. lactose
b. maltase

387

Other enzymes are named according to the substrate upon which they act. *Sucrase* is an enzyme that catalyzes the hydrolysis of sucrose.

a. Lactase is an enzyme that catalyzes the hydrolysis of ____________.

b. Which enzyme catalyzes the hydrolysis of maltose? ____________

COENZYMES

Some enzymes are proteins only. Other enzymes have a protein part and a nonprotein part, both of which must be present before the enzyme can function. The protein part of such an enzyme is called the *apoenzyme*; the nonprotein part is called the *coenzyme*.

$$\text{apoenzyme} + \text{coenzyme} \longrightarrow \text{enzyme}$$

Some vitamins act as coenzymes. Nicotinamide adenine dinucleotide, NAD, acts as a coenzyme in oxidation-reduction reactions in the mitochondria. Coenzyme A (CoA) functions in the metabolism of carbohydrates, lipids, and proteins.

388

a. The nonprotein part of an enzyme is called the ____________.

b. The protein part of an enzyme is called the ____________.

a. coenzyme
b. apoenzyme

389

One example of a coenzyme is ____________.

NAD or CoA or some vitamins

MODE OF ENZYME ACTIVITY

Enzymes are specific catalysts that contain an "active site," that section of the enzyme on which combination with the substrate takes place.

An older theory of enzyme activity, *the lock and key model*, assumes that the active site of an enzyme is rigid and that substrate molecules must fit into that site in order to react. Although it doesn't look like this in reality, one small part of it might be represented like this:

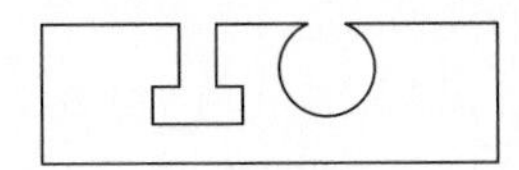

Enzyme

If two compounds, A and B, approach the enzyme, they can attach themselves to it, then react together and go off as a compound, AB. The enzyme is left free to react again and is not used up.

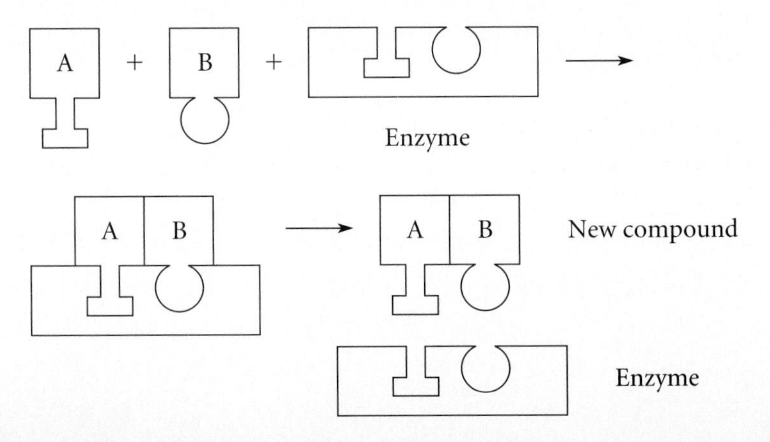

specific; increase

390

Enzymes are ___________ (specific/nonspecific) catalysts that ___________ the rate of chemical reaction in living systems.

A newer version of the mode of enzyme activity is the *induced fit theory*. In this theory the active site is flexible rather than rigid, and it changes its conformation upon binding to the substrate, thus yielding an enzyme-substrate combination.

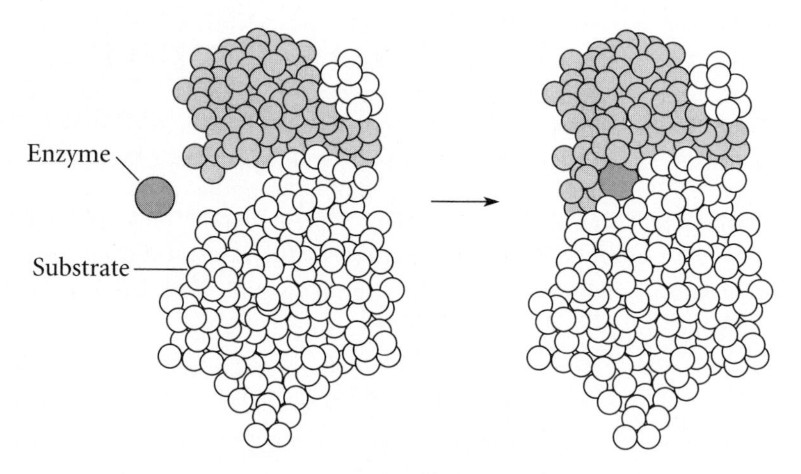

Induced fit model

a. rigid
b. flexible

391

a. In the lock and key model, the active site is ___________ (flexible/rigid).

b. In the induced fit model, the active site is ___________ (flexible/rigid).

INHIBITORS

392

If some substance other than the substrate fits into part of the active site, can the substrate then react with that enzyme? ____________

no

393

A substance that competes for a position in the active site of an enzyme is called a *competitive inhibitor*. Which of the following substances could act as a competitive inhibitor for the active site shown in problem 394?

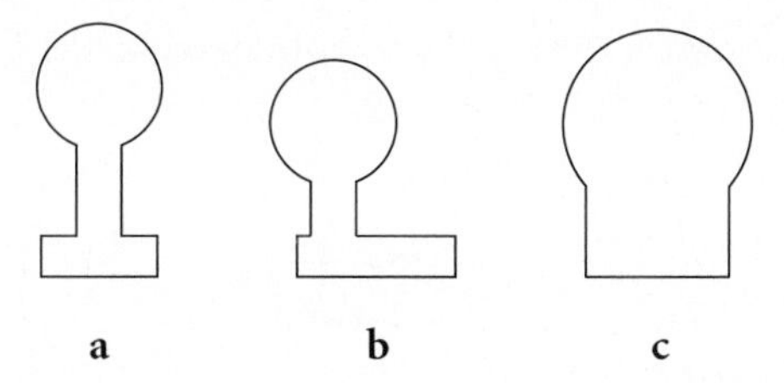

a

394

If substance **a** reacts with the active site of the enzyme shown here, could that enzyme then function normally? ____________

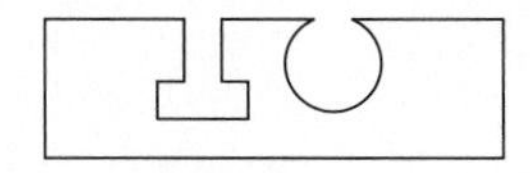

Enzyme

no

it should inhibit the growth of those bacteria

395

An example of a competitive inhibitor is sulfanilamide, whose structure is similar to that of p-aminobenzoic acid, a compound that is essential for the growth of certain bacteria.

SO_2NH_2 — benzene ring — NH_2

Sulfanilamide

$COOH$ — benzene ring — NH_2

p-Aminobenzoic acid

What effect should sulfanilamide, a competitive inhibitor, have on the growth of the bacteria that need p-aminobenzoic acid? ___________

no

396

Consider these diagrams:

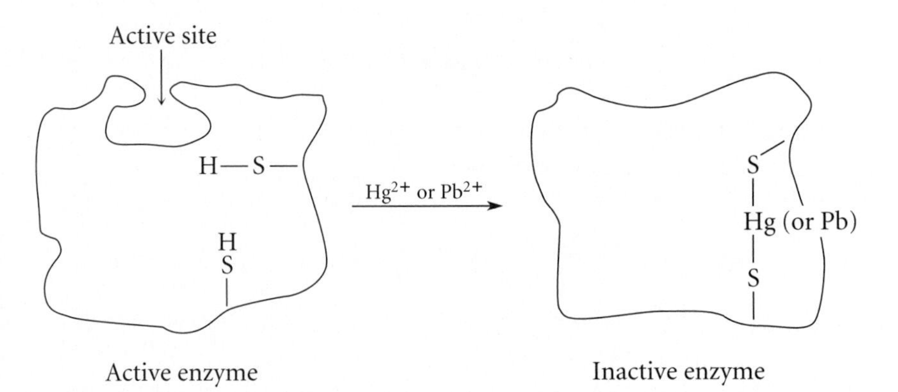

Note that the shape of the active site has been changed. Can the enzyme then function normally? ___________

397

A substance that changes the shape of the active site by reacting with the enzyme at some point other than the active site is called a *noncompetitive inhibitor.* What substance acts as a noncompetitive inhibitor in the reaction in problem 396? ___________

Hg^{2+} or Pb^{2+}

398

a. A substance that fits into the active site of an enzyme and prevents that enzyme from functioning normally is called a(n) ___________.

b. A substance that changes the shape of the active site of an enzyme by reacting with that enzyme at a position other than the active site is called a(n) ___________.

a. competitive inhibitor
b. noncompetitive inhibitor

Biologic Oxidation 15

Oxidation may be defined as:

- a loss of an electron (or electrons)
- a gain of oxygen
- a loss of hydrogen
- an increase in oxidation number

The compound lactic acid may be oxidized to pyruvic acid according to the following equation:

$$\mathrm{CH_3{-}CH(OH){-}COOH} \xrightarrow{\text{enzyme}} \mathrm{CH_3{-}CO{-}COOH} + \mathrm{H_2}$$

Lactic acid → Pyruvic acid

399

This is an oxidation reaction because it involves:

__ the gain of oxygen by the lactic acid molecule
__ the loss of hydrogen by the lactic acid molecule

the loss of hydrogen by the lactic acid molecule

a. alcohol and acid
b. ketone and acid

400

a. Lactic acid contains what functional groups? ____________

b. Pyruvic acid contains what functional groups? ____________

gain of oxygen

401

Another example of oxidation is the reaction of acetaldehyde to form acetic acid:

$$\underset{\text{Acetaldehyde}}{\mathrm{H{-}\overset{\displaystyle H}{\underset{\displaystyle H}{C}}{-}\overset{\displaystyle H}{C}{=}O}} \xrightarrow{[O]} \underset{\text{Acetic acid}}{\mathrm{H{-}\overset{\displaystyle H}{\underset{\displaystyle H}{C}}{-}\overset{\displaystyle OH}{C}{=}O}}$$

This reaction is oxidation because it involves ____________.

There is a loss of an electron.

402

The Fe^{2+} ion may be oxidized to the Fe^{3+} ion as follows:

$$Fe^{2+} - e^{-} \longrightarrow Fe^{3+}$$

Why is this oxidation? ____________

Reduction is the opposite of oxidation. Reduction is

- a gain of an electron (or electrons)
- a gain of hydrogen
- a loss of oxygen
- a decrease in oxidation number

Oxidation and reduction always occur together. One cannot occur without the other.

403

Consider the following reaction:

$$\begin{array}{c} COOH \\ | \\ H-C-H \\ | \\ H-C-H \\ | \\ COOH \end{array} \xrightarrow{FAD \quad FADH_2} \begin{array}{c} COOH \\ | \\ C-H \\ \| \\ C-H \\ | \\ COOH \end{array}$$

Succinic acid → Fumaric acid

a. Is succinic acid oxidized or reduced? ____________
   Why? ____________

b. Is FAD oxidized or reduced? ____________ Why? ____________

Note that both oxidation and reduction are involved in this reaction.

a. oxidized; there is a loss of hydrogen
b. reduced; there is a gain of hydrogen

404

Consider the following reaction:

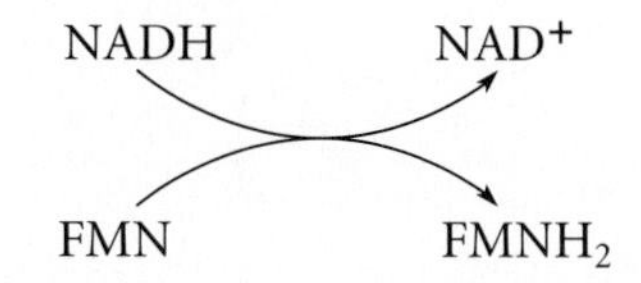

What was oxidized? ____________ Why? ____________

What was reduced? ____________ Why? ____________

NADH; it lost hydrogen; FMN; it gained hydrogen

CELLULAR RESPIRATION

Cellular respiration enables cells to produce energy from food through oxidation-reduction reactions.

Look at Figure A in the Appendix (page 210) and refer to it for problems 405–428. Note that cellular respiration involves three distinct stages: glycolysis, the Krebs cycle, and the electron transport chain.

cytoplasmic fluid; mitochondrion; mitochondrion

405

In which part of the cell does glycolysis take place? ____________

In which part of the cell does the Krebs cycle occur?____________

In which part of the cell does the electron transport chain function? ____________

pyruvic acid; cytoplasmic fluid

406

Glycolysis refers to the conversion of glucose to ____________ and takes place in which part of the cell? ____________

Krebs; mitochondrion

407

Pyruvic acid from the glycolysis sequence is changed to acetyl CoA and enters the ____________ cycle, which takes place in the ____________ of the cell.

NADH

408

What substance transfers electrons from the glycolysis sequence to the electron transport chain? ____________

409

Where does the NADH from the glycolysis sequence go? ____________

to the electron transport chain

410

Where else is NADH produced? ____________

in the Krebs cycle and in conversion of pyruvic acid to acetyl CoA

411

During glycolysis, glucose is converted to ____________.

pyruvic acid

412

Glycolysis takes place in the ____________ of the cell.

cytoplasmic fluid

413

In glycolysis, ____________ is converted to pyruvic acid.

glucose

414

Electrons are transferred from the glycolysis sequence to the __________ via what substance? ____________

electron transport chain; NADH

415

The electron transport chain functions in which part of the cell? ____________

mitochondrion

mitochondrion

416

The Krebs cycle functions in which part of the cell? ____________

3

417

Glycolysis produces pyruvic acid, NADH, and one other substance, ATP, the cell's main energy compound.

The abbreviation ATP refers to adenosine triphosphate. This name indicates that the compound adenosine is bonded to ____________ (how many?) phosphate groups?

2

418

The abbreviation ADP refers to adenosine disphosphate, indicating that the adenosine molecule is bonded to ____________ (how many?) phosphate groups?

adenosine; 1; phosphate

419

The abbreviation AMP refers to adenosine monophosphate and indicates that the ____________ molecule is bonded to ____________ (how many?) ____________ group(s)?

acetyl CoA

420

Pyruvic acid from the glycolysis sequence enters the Krebs cycle after being changed into ____________.

421

Electrons are transported from the Krebs cycle via ____________ and ____________.

NADH; $FADH_2$

422

Electrons are transported from the Krebs cycle to the ____________.

electron transport chain

423

The electron transport chain functions in which part of the cell? ____________

mitochondrion

424

NADH is produced in three different sequences during cellular respiration. They are ____________, ____________, and ____________.

glycolysis; Krebs cycle; conversion of pyruvic acid to acetyl CoA

425

$FADH_2$ is produced during which sequence? ____________

Krebs cycle

426

The Krebs cycle produces the energy compound ____________.

ATP

427

The electron transport chain produces the energy compound ________.

ATP

glycolysis; Krebs cycle; electron transport chain

428

During cellular respiration, ATP is produced in the following sequence: ____________, ____________, and ____________.

GLYCOLYSIS

Refer to Figure B in the Appendix (page 211) for problems 429–448.

glucose; 6

429

Glycolysis begins with ____________ (which compound?), which contains ____________ (how many?) carbon atoms?

pyruvic acid; 3

430

The end-product of glycolysis is ____________, which contains ____________ (how many?) carbon atoms?

2

431

For each molecule of glucose that goes through the glycolysis sequence, how many molecules of pyruvic acid are produced? ____________

cytoplasmic fluid

432

In which part of the cell does glycolysis take place? ____________

ADP

433

Note that in the first half of the glycolysis sequence ATP is used up and is converted to ____________.

434

How many ATPs are required for the first 3 steps in the glycolysis sequence? ____________

The ATP thus used is converted into ____________.

2; ADP

435

In the second half of the glycolysis sequence, energy is produced and ADP is converted to ____________.

ATP

436

In the second half of the glycolysis sequence, ____________ (how many?) ADPs are converted to ATPs.

4

437

In the first half of the glycolysis sequence, ____________ (how many?) ATPs are converted to ADPs.

2

438

In the entire glycolysis sequence, the net number of ATPs produced is ____________.

2

2

439

In addition to ATP, another energy compound, NADH, is produced during glycolysis. How many NADHs are produced from one molecule of glucose during the glycolysis sequence? ____________

NAD^+

440

NADH is produced from which compound? ____________

pyruvic acid

441

The overall glycolysis sequence may be written as:

glucose ⟶ ____________

2

442

The overall glycolysis sequence may be further written as:

glucose ⟶ ____________ (how many?) pyruvic acid

ATP

443

A further addition to this equation is:

2 ADP ____________

glucose ⟶ 2 pyruvic acid

444

This equation may then be modified to:

2 ADP ____ (how many?) ATP

glucose ⟶ 2 pyruvic acid

Answer: 2

445

This equation may be further modified to:

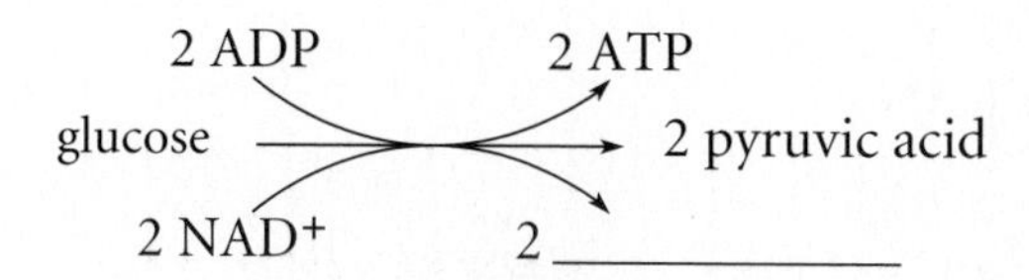

Answer: NADH

446

The complete glycolysis sequence may thus be written as:

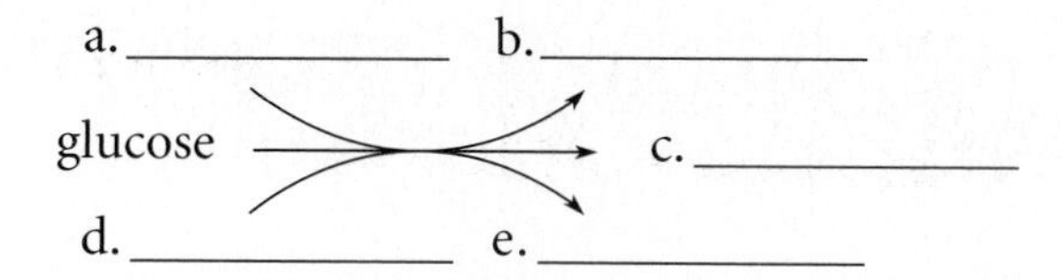

Answer:
a. 2 ADP
b. 2 ATP
c. 2 pyruvic acid
d. 2 NAD^+
e. 2 NADH

447

A biochemical reaction that requires oxygen is called *aerobic*; one that does not require oxygen is termed *anaerobic*.

Do any of the reactions in the glycolysis sequence require oxygen?

____

Answer: no

anaerobic

448

Is glycolysis an aerobic or an anaerobic sequence? ____________

CONVERSION OF PYRUVIC ACID TO ACETYL COENZYME A

Refer to Figure C in the Appendix (page 212) for problems 449–462.

pyruvic acid

449

The end-product of the glycolysis sequence is ____________.

cytoplasmic fluid

450

Pyruvic acid is produced in which part of the cell? ____________

mitochondrion

451

Pyruvic acid from the glycolysis sequence diffuses from the cytoplasmic fluid into which part of the cell? ____________

no

452

Does pyruvic acid enter the Krebs cycle directly? ____________

453

Pyruvic acid is converted into which compound in the mitochondrion?
____

acetyl CoA

454

Does acetyl CoA (acetyl coenzyme A) enter the Krebs cycle? __________

yes

455

Each molecule of glucose that enters the glycolysis sequence produces ____________ (how many?) molecules of pyruvic acid.

2

456

Each molecule of pyruvic acid that diffuses from the cytoplasmic fluid into the mitochondrion produces ____________ (how many?) molecules of acetyl CoA.

1

457

Therefore, during glycolysis, one molecule of glucose produces ____________ molecules of pyruvic acid, which in turn are used to produce ____________ molecules of acetyl CoA.

2; 2

6; 6; 2

458

One molecule of glucose contains ____________ (how many?) carbon atoms?

Two molecules of pyruvic acid contain ____________ (how many?) carbon atoms?

The acetyl groups of two molecules of acetyl CoA contain four carbon atoms.

How many carbon atoms are missing during this overall set of reactions?

____

2; pyruvic acid

459

As is shown in Figure C, in the cytoplasmic fluid, one molecule of carbon dioxide is produced for each molecule of pyruvic acid converted into acetyl CoA. Thus, one molecule of glucose enters the glycolysis sequence and is converted to ____________ (how many?) molecules of ____________.

2; acetyl CoA

460

In the mitochondrion, two molecules of pyruvic acid are converted into ____________ (how many?) molecules of ____________, which then enter the Krebs cycle.

carbon dioxide or CO_2

461

The missing carbon atoms show up as ____________.

462

For each molecule of acetyl CoA produced from pyruvic acid, ________ (how many?) molecules of NADH are produced. How many molecules of CO_2 are produced? ____________

1; 1

THE KREBS CYCLE

Refer to Figure D in the Appendix (page 213) for problems 463–473.

463

The two carbon atoms of the acetyl part of acetyl CoA participate in the Krebs cycle while the CoA molecule is recycled.

The two carbon atoms from the acetyl group are eliminated from the Krebs cycle as two molecules of what compound? ____________

CO_2

464

As one molecule of acetyl CoA goes through the Krebs cycle:

How many molecules of ATP are produced? ____________

How many molecules of NADH are produced? ____________

How many molecules of $FADH_2$ are produced? ____________

1; 3; 1

465

For every molecule of acetyl CoA that enters the Krebs cycle, _________ molecules of ATP, ___________ molecules of NADH, and ___________ molecules of $FADH_2$ are produced.

1; 3; 1

ADP

466

In the Krebs cycle, ATP is produced from what compound? ___________

NAD^+

467

In the Krebs cycle, NADH is produced from what compound? _________

FAD

468

In the Krebs cycle, $FADH_2$ is produced from what compound? _________

mitochondrion

469

In which part of the cell does the Krebs cycle take place? _____________

CO_2; ATP; NADH; $FADH_2$

470

The four products eliminated from the Krebs cycle are: _____________, _____________, _____________, and _____________.

no

471

Is oxygen required in any of the steps of the Krebs cycle? _____________

anaerobic

472

Is the Krebs cycle an aerobic or anaerobic sequence? _____________

473

Where do the NADH and the $FADH_2$ produced by the Krebs cycle go?
____

to the electron transport chain

ELECTRON TRANSPORT CHAIN

Refer to Figure E in the Appendix (page 214) for problems 474–485.

474

The reactions in the electron transport chain take place in which part of the cell? ___________

mitochondrion

475

Is oxygen required in any of the steps of the electron transport chain?
____

yes

476

Is the electron transport chain an aerobic or anerobic sequence?
____

aerobic

477

Which compounds enter the electron transport chain? ___________

NADH, $FADH_2$

478

The $FADH_2$ entering the electron transport chain comes from which sequence? ___________

Krebs cycle

glycolysis, conversion of pyruvic acid to acetyl CoA, and Krebs cycle

479

The NADH entering the electron transport chain comes from three different sequences. They are ____.

ATP

480

What is the primary energy compound produced by the electron transport chain? ____________

yes

481

Is ATP produced during glycolysis? ____________

no

482

Is ATP produced during the conversion of pyruvic acid to acetyl CoA? ____________

yes

483

Is ATP produced in the Krebs cycle? ____________

yes

484

Is ATP produced in the electron transport chain? ____________

ADP

485

ATP is produced from which compound? ____________

Review

486

What elements do the following symbols represent?

a. H ___________ b. O ___________ c. S ___________

d. Na ___________ e. K ___________ f. Fe ___________

a. hydrogen
b. oxygen
c. sulfur
d. sodium
e. potassium
f. iron

487

Write the symbol for each of the following elements:

a. chlorine _____ b. zinc _________ c. calcium _________

d. carbon ______ e. bromine _____ f. magnesium _____

a. Cl
b. Zn
c. Ca
d. C
e. Br
f. Mg

488

When an atom gains an electron, it forms an ion with what charge?

____

negative

489

The chlorine atom gains 1 electron to form a chloride ion. The symbol for the chloride ion is ____________.

Cl^{-}

ionic

490

When two ions are held together by the attraction of their opposite charges, what type of bond is between them? ____________

hydrogen (or H^+)

491

An acid is a substance that yields ____________ ions in solution.

hydrogen (or H^+)

492

A base is a substance that accepts ____________ ions in solution.

any ions except H^+ and OH^-

493

Salts yield what types of ions? ____________

an electrolyte

494

A solution that conducts electricity is called ____________.

a nonelectrolyte

495

A solution that does not conduct electricity is called ____________.

496

Which of the following are anions? ___________ Cations? ___________

Na^+ Cl^- Mg^{2+}
SO_4^{2-} K^+ NO_3^-

anions: Cl^-, SO_4^{2-}, NO_3^-; cations: Na^+, Mg^{2+}, K^+

497

Saliva has a pH between 5.5 and 6.9. What type of liquid is saliva?

____

a weak acid

498

Blood has a pH range of 7.35–7.45. What type of liquid is blood?

____

a weak base

499

An acid of pH 2.75 is how many times as strong as one of pH 3.75?

____

10

500

A solution that maintains a constant pH upon the addition of either acid or base is called a ___________.

buffer

sugar; water

501

In a solution of sugar in water, the solute is ____________ and the solvent is ____________.

yes; no

502

Do solutions pass through membranes? ____________ Do colloids? ____________

100

503

A 10% glucose solution will contain 10 g of glucose in ____________ mL of solution.

1 liter

504

A 1M KCl solution will contain 1 mole of KCl in how much solution? ____________

0.5M NaCl

505

Which solution will have a higher osmolarity, 0.5M NaCl or 0.5M glucose? ____________

covalent

506

When two atoms share electrons, they are held together by a __________ bond.

507

Diagram the structure of the hydrocarbon compound containing the following arrangement of carbon atoms, and indicate all the hydrogen atoms.

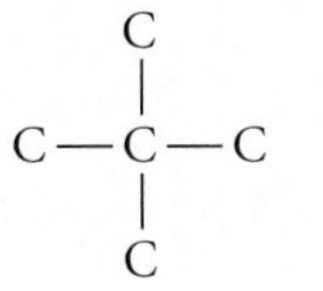

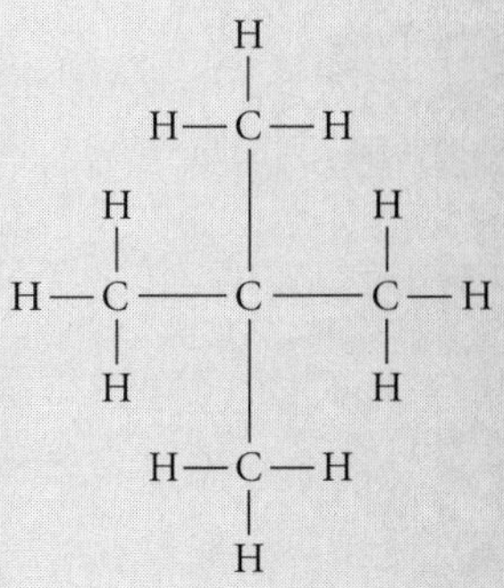

```
    C
    |
C — C — C
    |
    C
```

```
            H
            |
          H—C—H
            |
    H       |       H
    |       |       |
  H—C———————C———————C—H
    |       |       |
    H       |       H
            |
          H—C—H
            |
            H
```

508

Indicate the hydrogen atoms attached to the following arrangement of carbon atoms.

```
C≡C—C
```

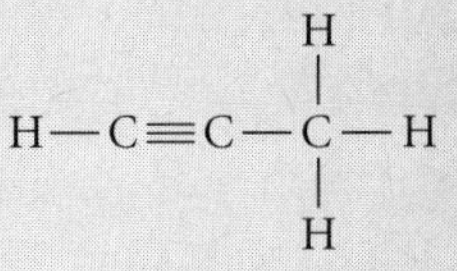

```
            H
            |
H—C≡C—C—H
            |
            H
```

509

What is the simplified structure of this compound?

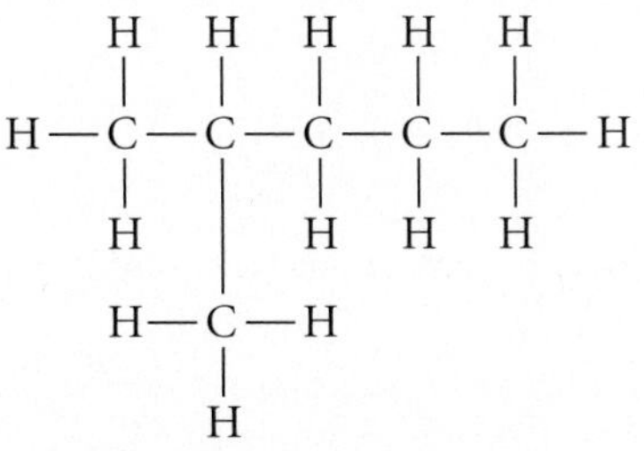

```
    H   H   H   H   H
    |   |   |   |   |
H — C — C — C — C — C — H
    |   |   |   |   |
    H   |   H   H   H
        |
    H — C — H
        |
        H
```

$CH_3CH(CH_3)CH_2CH_2CH_3$

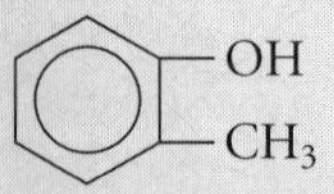

polar

hydrogen

hydrogen

510

What is the simplified structure of this compound?

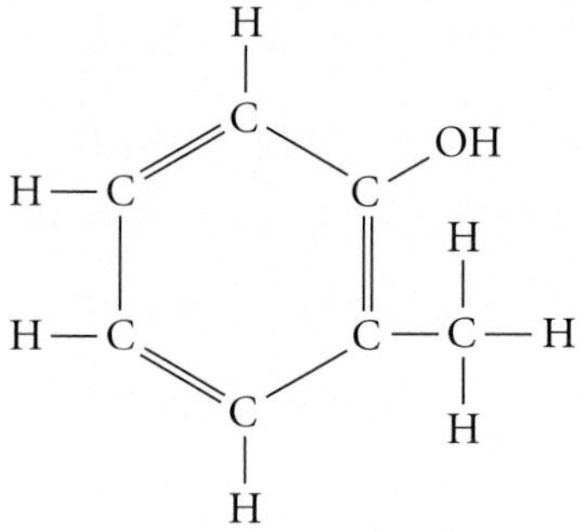

511

Is the water molecule polar or nonpolar? ____________

512

What types of bonds are present *between* water molecules? ____________

513

What type of bond is involved in maintaining the coil shape of a protein? ____________

514

When DNA replicates, it "unzips," breaking what type of bonds?
____

hydrogen

515

Are hydrogen bonds strong or weak? ___________

weak

516

In a protein, hydrogen bonds occur between an H attached to a(n) ___________ atom and an O that is part of a(n) ___________ group.

N; C=O

517

What type of functional group is represented by:

a. OH ___________
b. COOH ___________
c. C—C—C (with =O on the middle C) ___________
d. OPO_3H_2 ___________
e. NH_2 ___________
f. COOC ___________
g. CHO ___________

a. alcohol
b. acid
c. ketone
d. phosphate
e. amine
f. ester
g. aldehyde

Identify the types of functional groups present in each of the compounds in problems 518–520.

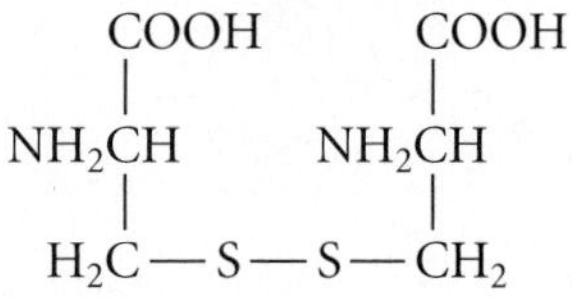

acid, amine, disulfide

518

$$\begin{array}{ccc} COOH & & COOH \\ | & & | \\ NH_2CH & & NH_2CH \\ | & & | \\ H_2C-S & - & S-CH_2 \end{array}$$

thiol or sulfhydryl

519

$CH_3CH_2CH_2SH$

a. amine and ketone
b. aldehyde and alcohols
c. alcohol and acid

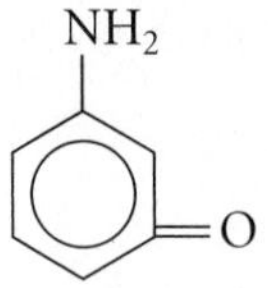

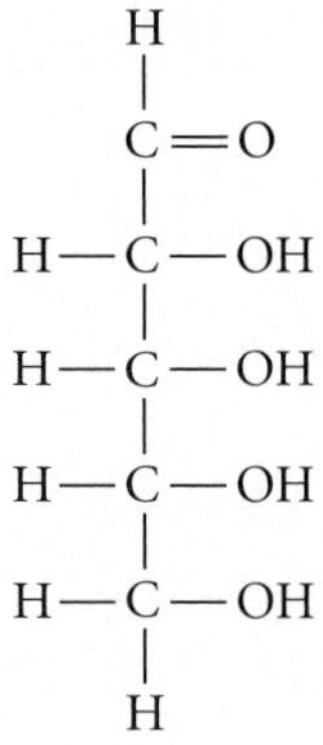

520

a. (benzene ring bearing NH_2 and $=O$)

b.
$$\begin{array}{c} H \\ | \\ C=O \\ | \\ H-C-OH \\ | \\ H-C-OH \\ | \\ H-C-OH \\ | \\ H-C-OH \\ | \\ H \end{array}$$

c. $CH_3CH(OH)COOH$

a. ____________ b. ____________ c. ____________

521

Which of the following compounds is a carbohydrate?

__ C_6H_6 __ $C_{12}H_{22}O_{11}$
__ C_6H_7ON __ C_2H_6O

$C_{12}H_{22}O_{11}$

522

a. Monosaccharides have names ending in ____________.

b. Most enzymes have names ending in ____________.

a. ***ose***
b. ***ase***

523

An example of a pentose is ____________; of a hexose is ____________.

ribose; glucose

524

The prefix *deoxy* means ____________.

with one less oxygen

525

Hydrolysis of a disaccharide yields ____________.

two monosaccharides

glycosidic

526

The bond holding the two halves of a disaccharide together is called a ____________ bond.

polysaccharide

527

Starch is a ____________ (monosaccharide/disaccharide/polysaccharide).

a. monosaccharides
b. polysaccharides

528

a. Which type of carbohydrate can be absorbed directly into the blood? ____________

b. Which type of carbohydrate forms a colloidal dispersion in water? ____________

oils

529

Which are unsaturated, fats or oils? ____________

fatty acids; glycerol

530

The products of fat hydrolysis are ____________ and ____________.

531

A phospholipid, on hydrolysis, yields ____________, ____________, ____________, and ____________.

fatty acids; glycerol; phosphoric acid; a nitrogen compound

532

An example of a steroid is ____________.

cholesterol or bile salts or sex hormones

533

Proteins always contain which elements? ____________

C, H, O, N

534

The hydrolysis of protein yields ____________.

amino acids

535

A polypeptide contains ____________ amino acids held together by ____________ bonds.

many; peptide

a; b and c; d; e

536

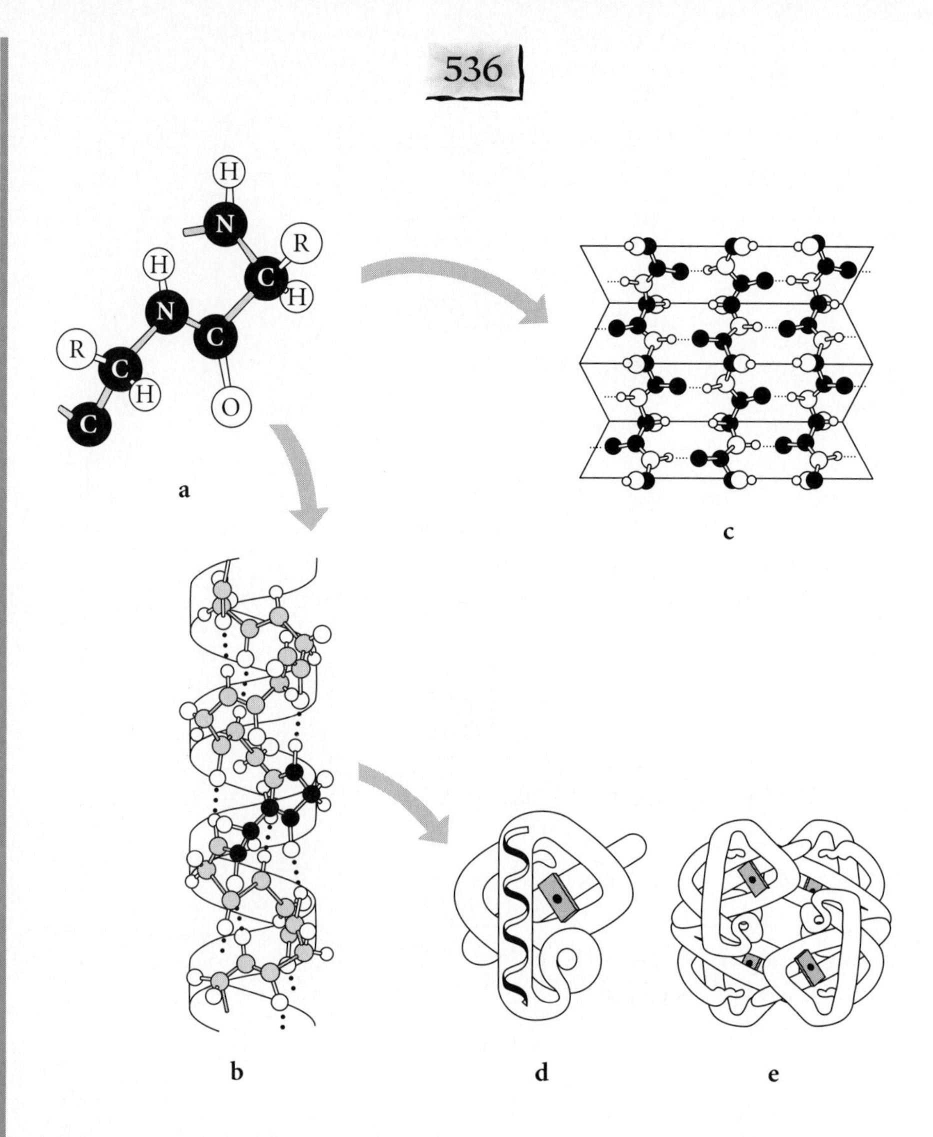

Which figure(s) represents the primary structure of a protein?

____

Which figure(s) represents the secondary structure of a protein?

____

Which figure(s) represents the tertiary structure of a protein?

____

Which figure(s) represents the quaternary structure of a protein?

____

537

A tripeptide consists of ____________ (how many?) amino acids linked together by what type of bond? ____________

3; amide or peptide

538

Nucleotides may contain one of two pentoses. They are ____________ and ____________.

ribose; deoxyribose

539

The compound that provides chemical energy for the body is ________.

ATP or adenosine triphosphate

540

ATP contains how many high-energy phosphate bonds? ____________

2

541

The opposite sides of the DNA coils are held together by ____________ bonds.

hydrogen

a. T
b. U
c. G
d. G

542

a. In DNA, the partner of A is ____________.

b. As RNA is synthesized, the presence of A in DNA will specify the incorporation of ____________ in the RNA.

c. As RNA is synthesized, the presence of C in DNA will specify the incorporation of ____________ in the RNA.

d. In DNA, the partner of C is ____________.

specific

543

Enzymes are ____________ (specific/nonspecific).

coenzyme

544

The nonprotein part of an enzyme is called the ____________.

a. competitive inhibitor
b. noncompetitive inhibitor

545

a. A substance that “fits” into the active site of an enzyme and prevents that enzyme from functioning is called a ____________.

b. A substance that alters the shape of the “active” site of an enzyme by reacting with that enzyme at a spot other than the active site is called a ____________.

546

Which of the two acids is oxidized? ____________

Which substance is reduced? ____________

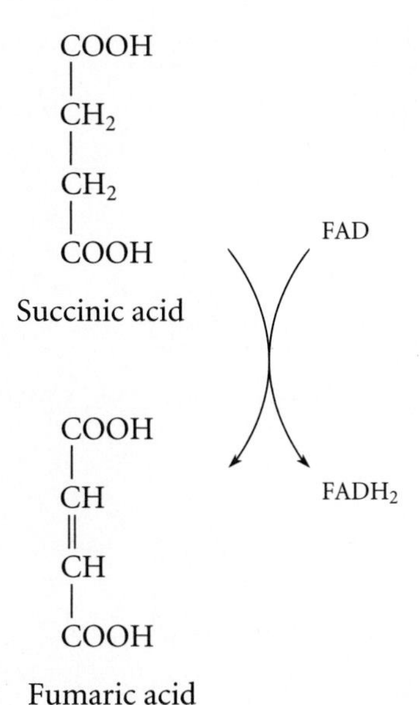

succinic acid; FAD

547

Glycolysis involves the conversion of glucose to ____________.

pyruvic acid

548

Glycolysis takes place in which part of the cell? ____________

cytoplasmic fluid

549

Is glycolysis an aerobic or anaerobic sequence? ____________

anaerobic

yes

550

Does glycolysis produce ATP? ____________

electron transport chain

551

The NADH produced during glycolysis is transferred to which sequence? ____________

acetyl CoA; mitochondrion

552

Pyruvic acid is converted into ____________ in which part of the cell? ____________

Krebs; mitochondrion

553

Acetyl CoA enters the ____________ cycle, which takes place in which part of the cell? ____________

anaerobic

554

Is the Krebs cycle an aerobic or anaerobic sequence? ____________

yes

555

Is ATP produced during the Krebs cycle? ____________

556

The NADH and $FADH_2$ produced during the Krebs cycle are transferred to which sequence? ____________

electron transport chain

557

The electron transport chain occurs in which part of the cell? _________

mitochondrion

558

Is the electron transport chain an aerobic or anaerobic sequence?

____

aerobic

559

In which parts of the cellular respiration sequence is ATP produced?

____

glycolysis, Krebs cycle, electron transport chain

Conclusion

Congratulations! You have completed the program. If you have worked conscientiously, you should now be able to:

- recognize the elements present in various biological compounds.
- understand the term *pH* as applied to fluids and cells.
- recognize what is meant by the terms *acids*, *bases*, and *salts* as they occur in plant and animal tissues.
- know what electrolytes are so that their functions in the life of the cell can be more clearly understood.
- understand oxidation and reduction as they occur in the metabolic processes of plants and animals.
- differentiate among carbohydrates, lipids, and proteins, and recognize each from its unique functional groups.
- understand how enzymes function.
- know what nucleic acids are and how they are held together.

Appendix

Figure A
Cellular Respiration

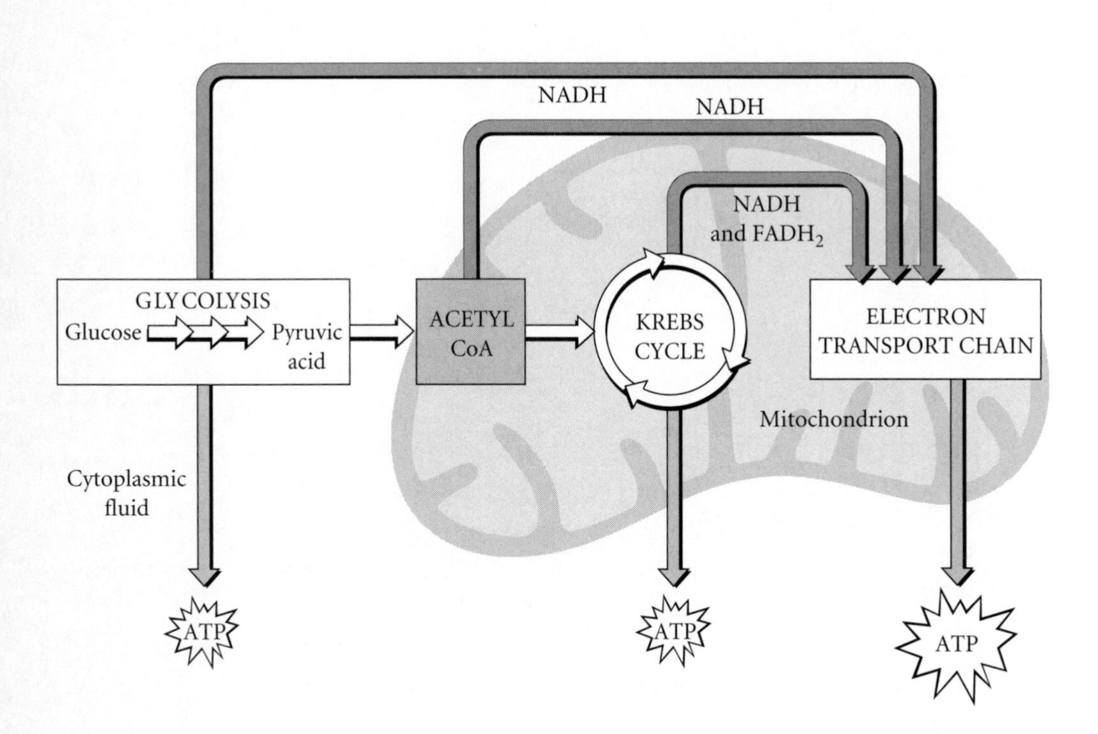

Figure B
Glycolysis

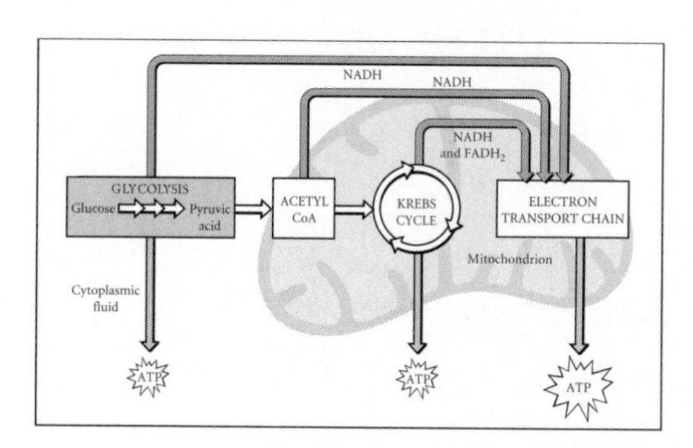

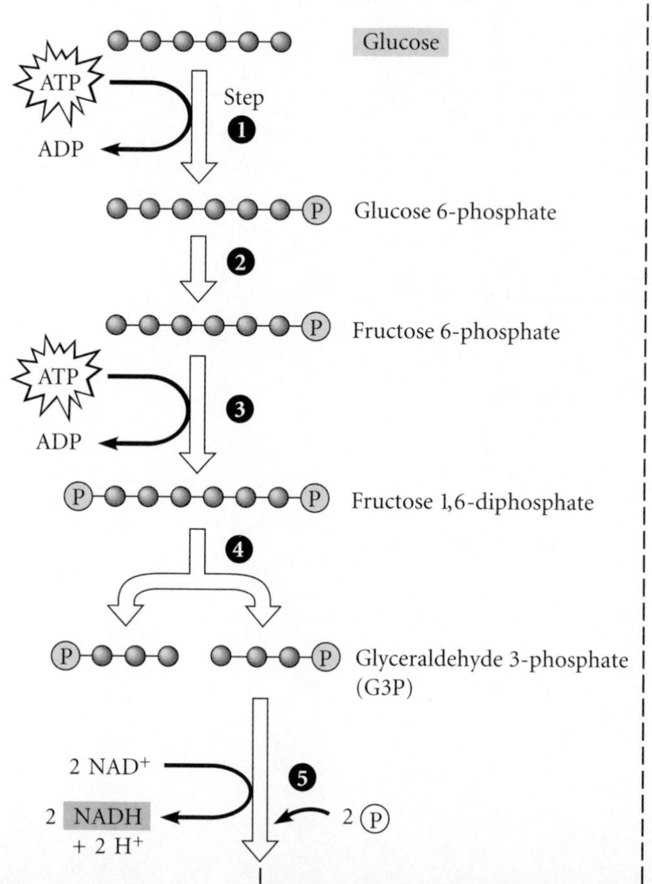

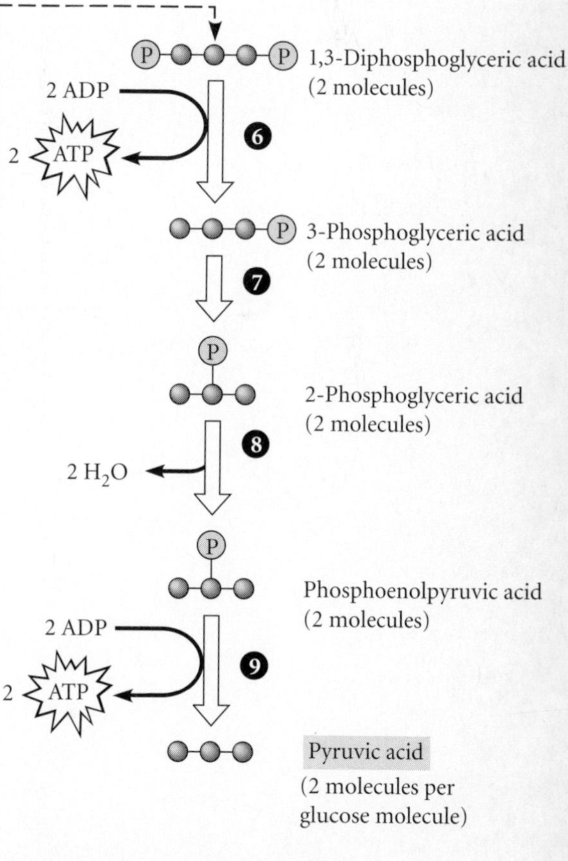

Figure C
Conversion of Pyruvic Acid to Acetyl Coenzyme A

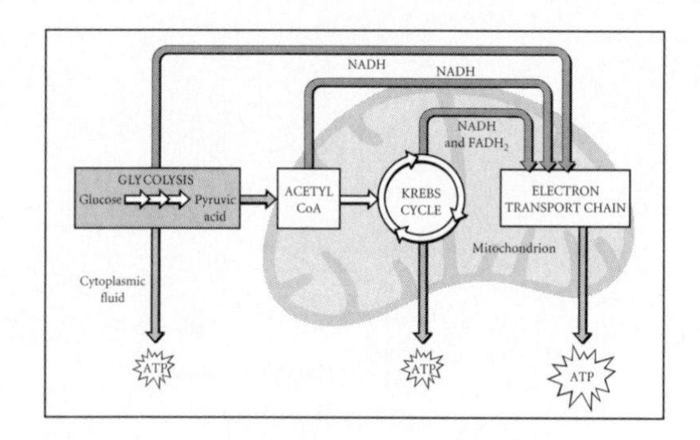

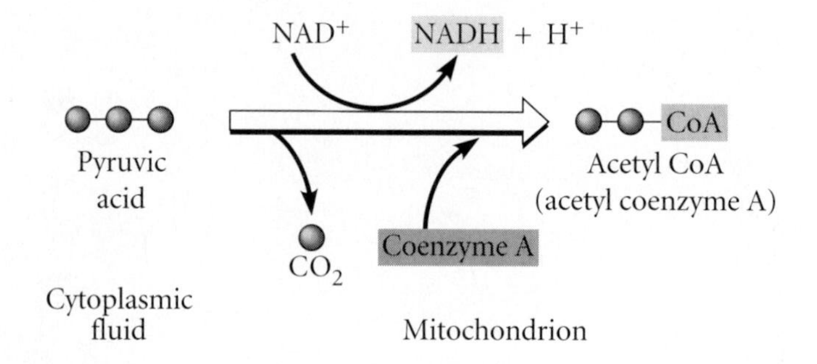

Figure D
The Krebs Cycle

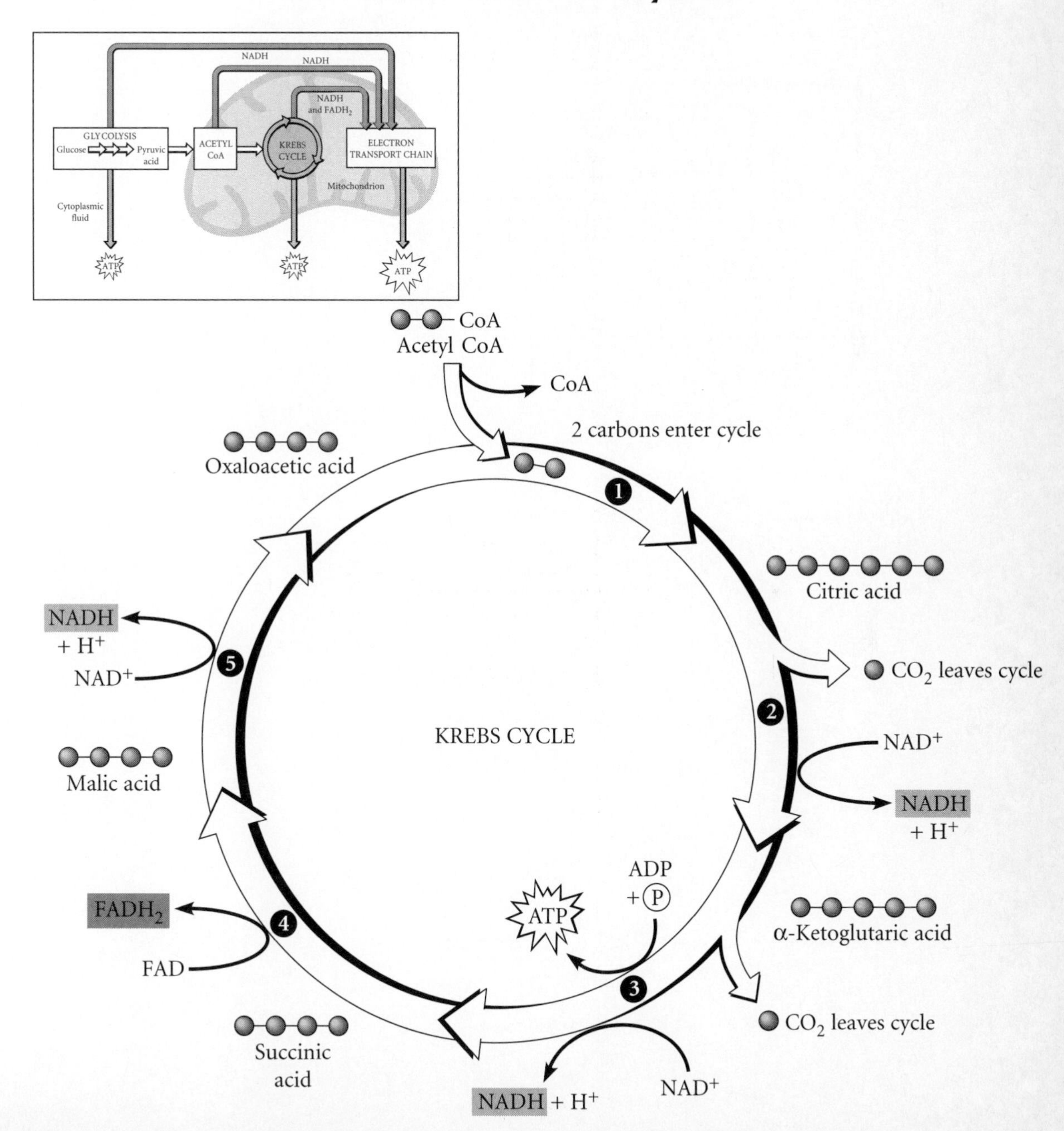

Figure E
Electron Transport Chain

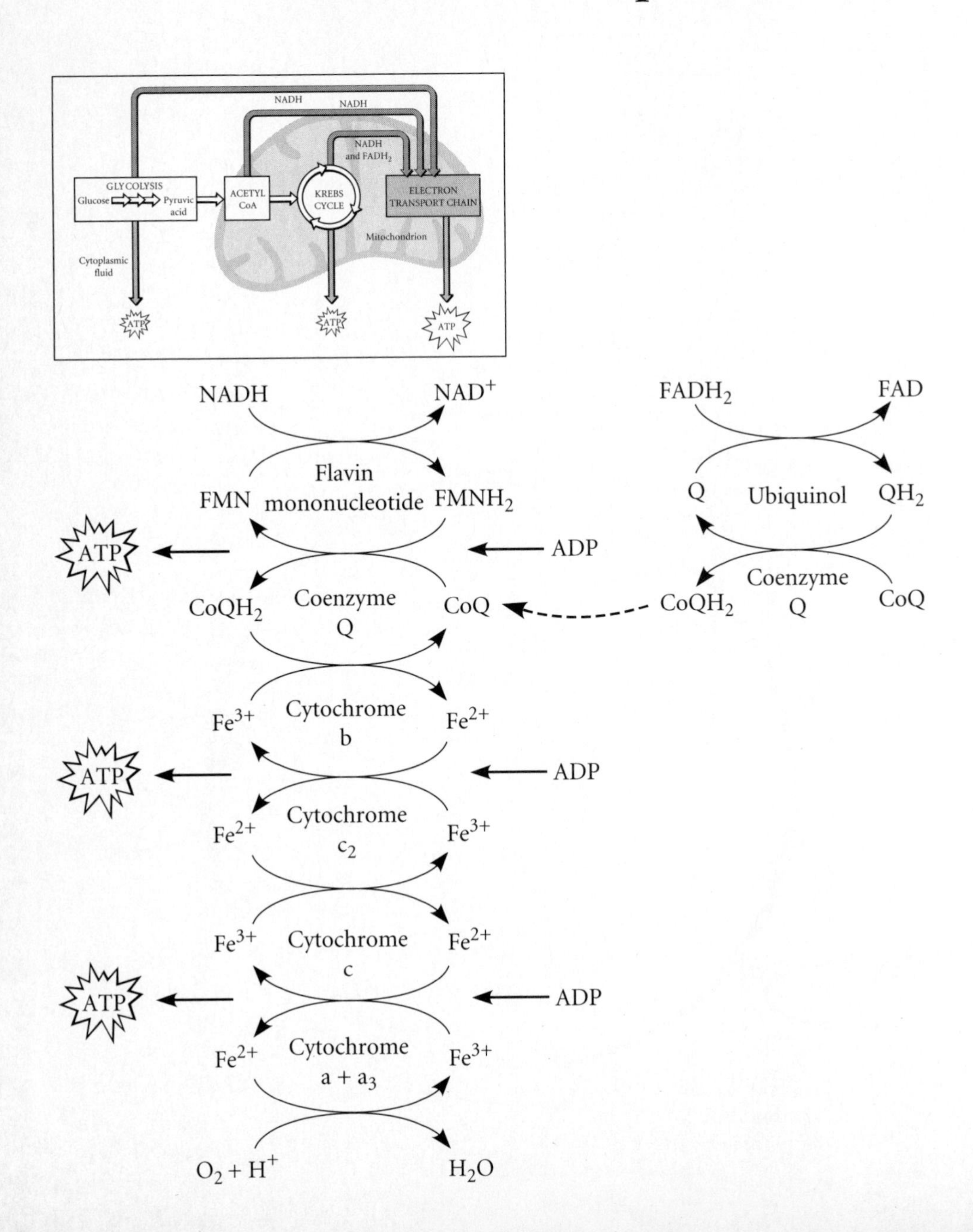